The Intermediate Licence Manual
for Radio Amateurs

3rd Edition

by

Roger Bleaney, M0RBK

with contributions by
George Smart, M1GEO
David Mills, G7UVW

Published by
Radio Society of Great Britain of 3 Abbey Court, Priory Business Park, Bedford MK44 3WH, United Kingdom
www.rsgb.org

1st Edition printed 2020 & 2021

2nd Edition printed 2021, 2021 & 2023

This 3rd Edition first printed 2024

Reprinted 2024

© Radio Society of Great Britain, 2024. All rights reserved. No part of this publication may be reproduced, stored in a retrieval system, or transmitted, in any form or by any means, electronic, mechanical, photocopying, recording or otherwise, without the prior written permission or the Radio Society of Great Britain.

ISBN: 9781 9139 9557 7

Cover design: Kevin Williams, M6CYB
Editing: Roger Bleaney, M0RBK
Production & revisions: Mark Allgar, M1MPA
Typography and design: Mark Pressland

The opinions expressed in this book are those of the author and are not necessarily those of the Radio Society of Great Britain. Whilst the information presented is believed to be correct, the publishers and their agents cannot accept responsibility for consequences arising from any inaccuracies or omissions.

Printed in Great Britain by the Short Run Press Ltd of Exeter, Devon

Any amendments or updates to this book can be found at: www.rsgb.org/booksextra

Radio Society of Great Britain
Advancing amateur radio since 1913

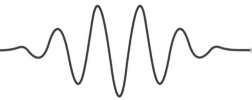

JOIN US TODAY
and get the best amateur radio magazine for

12 EDITIONS FOR ONLY £29.99

The Radio Society of Great Britain (RSGB) is run by radio amateurs for radio amateurs (licenced or not), working for you, protecting your interests. We keep you informed of the latest amateur radio news, and amongst friends who understand the hobby.

Being part of the RSGB means that we post direct to your door each month the biggest and best amateur radio magazine, *RadCom*. Only available to RSGB members for less than the price of some other high street radio magazines, there is no better way to stay in touch with the world of radio. You also receive its sister e-publications *RadCom Basics* and *RadCom Plus*.

Being a member of the RSGB is much more than a subscription to our magazines. You become part of a society that also provides all the great benefits shown overleaf.

If you want to get the most out of amateur radio, there is simply no better way to do that than by joining the Radio Society of Great Britain

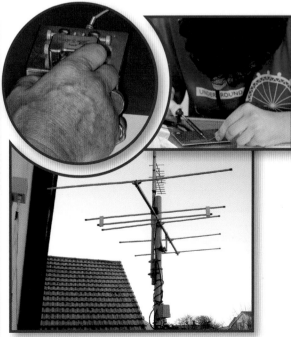

The Radio Society of Great Britain is a company limited by guarantee. Registered in England & Wales No. 216431
Registered Office: 3 Abbey Court, Fraser Road, Priory Business Park, Bedford, MK44 3WH, Tel: 01234 832700

You become part of the society that provides

The RSGB Regional Team:
Over 50 regional representatives who volunteer to help their fellow amateurs. Providing practical help and advice your local Regional Team also feeds back your concerns into the RSGB and onwards to National and International level.

Protection of your Hobby:
Run by radio amateurs, the RSGB protects your interests both nationally and internationally. We negotiate with the government on your behalf to ensure the future of our spectrum and the interests of the hobby.

RSGB Members Only Web Content:
Web pages exclusively for Members, containing videos, Radcom archive material, Bletchley Park free admission voucher and much more.

Book Discounts:
RSGB Members receive at least a 15% discount on the Society's extensive list of some 200 amateur radio books, CDs, RSGB branded merchandise, etc.

GREAT BENEFITS!
- RADCOM
- RADCOM PLUS AND BASICS
- QSL BUREAU
- MEMBERS ONLY WEB CONTENT
- BLETCHLEY PARK TICKETS
- BOOK DISCOUNTS
- RSGB REGIONAL TEAM HELP
- PROTECTION OF YOUR HOBBY
- RSGB CONTESTS AND AWARDS
- PLANNING ADVICE
- EMC ADVICE, AND MORE

Only £29.99 for a year! Join Today!

As an introduction, all newcomers into the exciting world of amateur radio can become a member of the Radio Society of Great Britain for the amazing price of £29.99 (paying by direct debit) for a full year's membership. This is a saving of nearly 60% on the price of a full membership. If you are a licenced amateur under 21, membership of the RSGB is free. Simply use the form below sending proof of your age and details of your licence.

Please note that we only accept original forms, in conjunction with the sale of this book. This offer is not available to anyone who has been a member of the RSGB before.

INTERMEDIATE LICENCE - OFFER

Please fill in section 1 and either 2 or 2a (under 21's section 1 only)

① Personal details
- Callsign (if any) _____ Initials _____
- First Name _____
- Last Name _____
- Address _____
- Postcode _____ Date of Birth __/__/__
- Tel No _____
- Email _____

② Payment alternatives
☐ I enclose a cheque for the sum of **£34.99** (£5.00 more than DD)
(Please make Cheques/PO (drawn in £ sterling only) payable to the "Radio Society of Great Britain".)

☐ Please debit £34.99 to my: VISA / Mastercard / Delta / Switch
Credit Card Details: Card No
☐☐☐☐ ☐☐☐☐ ☐☐☐☐ ☐☐☐☐

Card Expiry _____ Valid from _____
ISSUE No (Switch etc) _____ CW2 No ☐☐☐
Signature _____ Date _____

or ②ⓐ Direct debit instructions (Save on the cash price by paying by direct debit)

Instruction to your Bank or Building Society to Pay Direct Debit
☐ Annually ☐ Quarterly ☐ Monthly (please tick)   Service user number: 9 4 1 3 0 2

Please note: £29.99 is taken for all memberships at the outset. Monthly and quarterly DDs begin on the second year of membership.

1. Name of your Bank or Building Society Branch _____
2. Bank or Building Society Account number ☐☐☐☐☐☐☐☐   3. Branch Sort Code ☐☐☐☐☐☐
   (from the top right hand corner of your cheque)
4. Instruction to your Bank or Building Society
   Please pay Radio Society of Great Britain Direct Debits from the account detailed in this Instruction subject to the safeguards assured by the Direct Debit Guarantee.

Signature _____ Date _____

Please complete this form and send it to RSGB, 3 Abbey Court, Fraser Road, Priory Business Park, Bedford MK44 3WH

| FOR SOCIETY USE ONLY | RSGB Direct Debit Ref No. |

③ Extra services Do you wish to sign up for our online magazines and services too?
- ☐ Radcom Basics - for the less expert.
- ☐ Radcom Plus - for those who want more.
- ☐ Membership Benefits - discount vouchers & special offers.
- ☐ GB2RS - RSGB weekly news emailed to you
- ☐ Online Event Notifications

NB. The RSGB does not sell its data to third parties and information will only ever be sent by the RSGB

PLEASE RETAIN

- If an error is made in the payment of your Direct Debit, by Radio Society of Great Britain or your bank or building society you are entitled to a full and immediate refund of the amount paid from your bank or building society
- If you receive a refund you are not entitled to, you must pay it back when Radio Society of Great Britain asks you to
- You can cancel a Direct Debit at any time by simply contacting your bank or building society. Written confirmation may be required. Please also notify us.

DIRECT Debit Guarantee
- This Guarantee is offered by all banks and building societies that accept instructions to pay Direct Debits
- If there are any changes to the amount, date or frequency of your Direct Debit Radio Society of Great Britain will notify you 7 working days in advance of your account being debited or as otherwise agreed. If you request Radio Society of Great Britain to collect a payment, confirmation of the amount and date will be given to you at the time of the request.

Sinotel UK Limited
Unit 1 Block B
Harriott Drive
WARWICK
CV34 6TJ

SALES LINE: 01926-46020
Website: www.sinotel.co.uk E-Mail: sales@sinotel.co.uk

We also stock a wide range of antennas, batteries, programming cables, connecting leads, speaker microphones and adapters. We are authorised distributors for all the products that we sell.

BAOFENG T57
- 2m/70cm dual bander
- 5 Watt output
- IP57 Ingress rating
- Drop-in charger
- LED torch
- 1750 Hz Burst tone
- Earpiece/microphone
- 1800 mAh Li-ion battery

£39.99

XIEGU X5105 5 WATT HF RIG
- Ideal for portable QRP operation
- 5 Watt output
- Built-in ATU & 3800 mAh battery
- AM/CW/FM/SSB modes
- Robust aluminium housing
- SDR Technology
- As reviewed in RadCom 11/2018

£32.99

BAOFENG UV-3R
- Pocket-sized handheld
- 2m & 70 cm coverage
- Ideal Foundation radio
- LED torch
- Drop-in charger
- 2 Watt output (VHF/UHF)
- 1400 mAh Li-ion battery
- PC programmable

£24.99

XIEGU G90 20 WATT HF RIG
- 0.5 to 30 MHz (incl. WARC bands)
- 0-20 Watt variable output
- Full colour TFT Screen
- Multifunction microphone
- Detachable front panel
- 0.5 ppm TCXO
- S/Power/SWR meter
- AM/CW/SSB modes
- Spectrum/waterfall display
- Built-in ATU
- SO239 antenna connector
- 12-14 V DC supply required
- SDR Technology

£TBA

NEW!

VERO VR-6600PRO
- 2m/70cm remote head mobile
- 50 Watt output
- Wideband receive
- Large highly visible display
- Switchable display colours
- 1000 channel memory
- Full duplex operation
- 12-14 V DC supply required

£249.99

NEW!

TYT TH-UVF9
- 2m/70cm handheld
- 5 Watt output
- Highly efficient antenna
- OLED display
- Car charger included
- 1600 mAh Li-ion battery
- As reviewed in RadCom

£37.99

XIEGU G1M
- Mini HF QRP transceiver
- 8 Watt output
- Covers 5 HF bands
- CW/SSB
- 12-14 V DC supply required
- Ideal for Foundation licencees

NEW!

£TBA

T101 VHF/UHF ANTENNA ANALYSER
- Rugged Plastic Housing
- PC control software included
- 100-170/400-470 MHz
- N female test port
- Requires 2 x AA batteries
- Measure SWR/Impedance/Reactance/Return Loss

£189.99

TYT TH-8600
- 2m/70cm dual-band mobile
- 25/20 Watt output
- Super compact size - 110x140x42mm
- 200 memory channels
- Twin VFOs
- Windows PC programmable with included programming cable
- 12-14 V DC supply required

NEW!

Authorised UK distributors for

PLEASE VISIT OUR WEBSITE FOR THE LATEST NEWS & SPECIAL OFFERS

Errors & omissions excepted. All items subject to availability. Prices do not include carriage. All prices & specifications subject to change without notice.

Contents

Preface	iv
About the Authors	iv
Introduction	v
Syllabus v1.6 Cross Reference	vi
1: Licence Conditions	1
2: Operating Techniques	4
3: Tools, Construction & Safe Practice	7
4: Basic Electronics	12
5: Semiconductors	27
6: RF Oscillators	32
7: Transmitters & Receivers	34
8: Good Radio Housekeeping	45
9: Harmonics and Spurious Emissions	48
10: Antenna Matching	50
11: Feeders and Baluns	54
12: Antenna Concepts	58
13: Propagation	61
14: Measurements	65
15: The Examination	67
Appendix 1 - The relationship between power and voltage	68
Appendix 2 - Protective Multiple Earthing	68
Appendix 3 - Scientific Notation	69
Appendix 4 - Formulae from RSGB EX308	70
Appendix 5 - Experimentation for Learning	71

Preface

In 1990, the Novice Amateur Radio Licence was introduced as a way of attracting newcomers into the hobby. The Novice Amateur Radio Licence eventually morphed into the Intermediate licence, alongside the creation of the Foundation licence in 2002.

A new book called *Intermediate Licence - Building on the Foundation* was created for the new three-tier exam structure written by Steve Hartley, G0FUW with assistance from Dr John Craig, G3SGR, Ed Taylor, G3SQX and Alan Betts, G0HIQ. This book drew heavily on *The Novice Licence Student's Handbook* by John Case, GW4HWR which had been created to support the original syllabus. The new book ran to six editions containing the minor revisions as the syllabus bedded in and later larger changes in Licence Conditions and Operating Procedures. With the fifth edition there were significant changes in technical content requiring a greater understanding in order to justify the privileges of the Intermediate Licence, as required by Ofcom. The final sixth edition of the book added more detail around handling very large and very small numbers.

This book was replaced by this book *The Intermediate Licence Manual* which has now run to three different editions. This third edition is designed to meet syllabus version 1.6, which adds material on the new licence specifications introduced in 2024 and builds on the 2019 syllabus changes which repositioned the Intermediate level appropriately between the Foundation and Full licence syllabuses.

Writing this manual would not have been possible without help. The author wishes to acknowledge the hard work, dedication and patience by all concerned in the production of the previous books. In particular, thanks go to Dr George Smart, M1GEO, and Dr David Mills, G7UVW for their major contribution to the first edition of this book. The author is grateful for the professional input of the RSGB Staff. Finally, thanks go to Alan Betts, G0HIQ for his constant support and enthusiasm. On behalf of the entire amateur community I wish you every success and happiness in your Intermediate Licence studies.

Roger Bleaney, M0RBK

About the Authors

The authors, Drs Roger Bleaney, George Smart, and David Mills have extensive experience in the teaching and development of amateur radio through their work in clubs, involvement with the RSGB and membership of the Examinations Syllabus and Review Group (ESRG). They have been keen operators for many years.

Roger has been licensed since 2012 as M0RBK. He did not join the amateur radio community until retirement. His working life had been spent in research in the chemical industry and university followed by teaching science in the classroom. As a self-confessed radio addict he enjoys all aspects of the hobby, particularly, teaching radio theory and showing how the physical sciences have contributed to its evolution. It has been his pleasure to use the extra time given by retirement to continue the development of this Intermediate Licence Manual to this second edition.

Lead author of the first edition of this book George became licensed in 2002 as 2E1ZZZ with a Novice Amateur Radio Licence before progressing on to the Full licence in 2003. He is actively involved in training with the Loughton and Epping Forest Amateur Radio Society amongst other clubs, as well as being regularly active in contests and field events with the Secret Nuclear Bunker Contest Group and the Camb-Hams. He currently works in RF IP design & validation.

David was first licensed in 1994 as 2E1CZR and later became G7UVW. He has been interested in radio and electronics since he was old enough to wield a screwdriver. He is an active member of the Secret Nuclear Bunker Contest Group. He works at Queen Mary University of London, where he spends his days designing and building CT scanners and microscopes. David is also an active contester.

Introduction

Preparing for the Intermediate Licence

Welcome to the next stage in your self-training in amateur radio! The fact that you are reading this suggests that you have already passed the Foundation Licence examination. You may have been operating with your Foundation callsign for some time. No matter how long it is since you first became interested in amateur radio, this book is intended to prepare you for the Intermediate Licence exam.

You may not have operated with a Foundation Licence, but you must have passed the Foundation examination before you can take the Intermediate examination. There is no requirement for you to attend a formal training course to prepare for the examination, but it is without doubt the best way to learn if you can do so. If you cannot get to a local course, there is no need to worry. This book has been written to enable you to learn on your own before taking the examination. You may find it useful to review the Foundation book if it has been some time since you passed.

Your examination can be booked and completed online. If you are studying alone you can book your examination when you are ready to take it at, www.rsgb.org/exampay. The RSGB website, www.rsgb.org will also give information on contacting local clubs and lots of useful examination materials.

The Intermediate Specification

The Intermediate specification covers many of the same topics as at Foundation level, so you should be reasonably familiar with the basics. At this level you will be building on your Foundation knowledge to gain a better understanding of radio theory, improving your knowledge of operating techniques and learning about new concepts.

Appendix 6; Experimentation for learning, gives you details of projects you might like to try for yourself. They have been designed to bring aspects of the theory to life and can form the basis of discussion with other amateurs and club members.

Using this book

The chapter titles of this book try to match the sections in the Intermediate syllabus where possible. However, the chapters have been reordered slightly to set out a more logical scheme of work for you to follow.

The book is presented as a series of chapters which aim to walk you through the syllabus points. These can be read as many times as you require. You should take the time to understand the content in each section. If you find yourself struggling to understand a topic, try speaking with the tutors at your local radio club, or a fellow amateur who can advise you. There is also a wealth of information online, and many excellent video tutorials published on YouTube.

Guidance on the examination is given in Chapter 15: The Examination.

It is not possible to say how long study for the Intermediate licence should take since the time for you to complete this course of study depends on your level of knowledge and your experience. However, the only prior knowledge you need is that which you gained from preparing for the Foundation Licence assessments.

Other Resources

This book contains all the information you will need to prepare for the Intermediate Licence examination. However, there are lots of other sources of information available, for example:
- The Intermediate Licence terms and conditions are set out in the licence document available from Ofcom – all examinable parts are included in this book.
- The Radio Society of Great Britain's bookshop has a host of amateur radio books available.
- The RSGB website is a valuable source of material to support your studies. You can download, the reference booklet used in the examination (EX308), information about on-line training, the Exam Specification, sample examination questions and Mock Examination Papers
- Online forums and groups can provide a place to ask questions
- Video sharing platforms such as YouTube can help master practical skills
- The examination objective numbers are quoted in a cross-referenced index provided on page vi of this book. These will help you track down the learning material that supports each examination objective. Where additional resources are considered to be particularly helpful, details are included in the text.

After the Exam

A pass in the Intermediate examination will let you apply for an Amateur Radio (Intermediate) Licence. This will allow you to increase your output power and give you access to more bands than your Foundation Licence.

Online applications enable you to obtain your new Licence free of charge. You can apply by post if you wish, but there is a charge for paper applications.

You can make your application as soon as RSGB headquarters have confirmed that you have passed the exam and entered it into the Ofcom database. This takes just over one week, and so RSGB Head Quarters should not be contacted about confirmation of your result for at least ten days.

The next question is when to take the next step towards gaining a Full licence. You may wish to build on your success straight away, or you may prefer to use your new callsign for a while. It is entirely up to you, but we wish you success whichever path you choose.

Syllabus v1.6 Cross Reference

Syllabus No.	Page No.
Section 1	
Licensing conditions and station identification	
1A2	1
1B1	1
1C1	1
1C2	1
1D1	2
1D2	2
1E1	2
1F1	2
1G1	2
1H1	3
Section 2	
Technical aspects	
2A1	14,17,19,31
2C1	13,14
2C2	15,16
2C3	16
2D1	16,17,18
2D2	17,18
2D3	16
2D4	18,19,23
2D5	19
2D6	23
2E1	22
2E2	12,19,20,21,22
2E3	20,22,23,24
2E4	23,24
2E5	19,24
2E6	19,22,24
2E7	21
2E8	54
2F1	43
2G1	24
2H1	25
2H2	25,26
2H3	25
2H4	39
2H5	26
2I1	27,28
2I2	28,37
2I3	30,31
2I4	31
2I5	31,32
2I6	30
2J1	16
2J2	28,29
2J3	28,29,30
2J4	29,30
Section 3	
Transmitters and receivers	
3A2	36,37
3A3	37
3B1	35,36
3C1	32,33
3C2	33
3C3	33
3E1	36,37
3E2	37
3E3	37
3F1	38
3G2	38
3G3	26
3G4	35
3G5	35
3H2	38,39,40,41,42
3H3	39
3H4	38,39
3I1	40,41
3I2	41
3I3	41
3K1	38,41
3L1	42
3M1	43,44
3M2	44
3M3	43
Section 4	
Feeders and antennas	
4A1	55
4A2	55,56
4A3	54,55,56,57
4B1	57
4C2	59
4C3	58
4C4	59
4C5	51,60
4D1	58
4D2	60
4E1	52,53
4F1	53
4H1	78
Section 5	
Propagation	
5A2	61,64
5A3	61
5A4	60
5B1	61,62
5B2	62,63
5B3	62,63,64
5B4	62,63
5B5	62
5C3	63,64
Section 6	
Electro magnetic compatibility (EMC)	
6A1	47
6A2	47
6A3	11
6A4	47
6B1	47
6B2	47
6B3	47
6C1	47
6C2	47
6D1	47
6D2	51,52
6D3	75,76
6D4	48,50
6E1	34,45,46
6E2	46
6E3	11
6F2	45
6F3	45
Section 7	
Operating practices and procedures	
7A3	6
7A4	6
7B1	4,6
7E1	4
7F2	5,6
7G1	5
7G2	5
7G3	5
7G4	5
Section 8	
Safety	
8A1	17
8A4	9,10
8A6	9
8A8	16
8B2	8
8B3	8
8B4	8,9
8B5	9
8B6	8,9
8E1	11
Section 9	
Measurements and construction	
9A1	66
9A2	65,66
9A3	65
9A5	65
9B1	56,57,58
9C1	14
9D1	9
9E1	7
9E2	7
9E3	7
9E4	7

1: Licence Conditions

Licence Material included in the Examination

1A2 Licensing

Recall the restrictions applicable to Intermediate licensees in operation from a ship or aircraft.

The UK amateur radio licence authorises use of the designated spectrum across and over the UK and the Crown Dependencies of Jersey, Guernsey and the Isle of Man, including their territorial seas. This covers the area out to twelve nautical miles or until a point is reached halfway to another country, such as France. The licence also applies to other areas covered by UK law such as UK controlled areas of the North Sea.

The licence continues to apply if used on a UK registered ship or aircraft in international waters or airspace. The licence does not cover the use on a non-UK registered ship or aircraft internationally nor any use within the territorial waters or airspace of another country.

The licence also requires you to obtain the permission of the captain (or the person in charge at that time) of the ship or aircraft before transmitting and comply with any instructions they give, including stopping transmitting.

Airborne operation must be restricted to 500mW EIRP in primary amateur radio bands only, as detailed in Schedule 1 of the Amateur Conditions Booklet.

1B1 Supervision, Radio Equipment and use of correct Callsign

Understand the meaning of direct supervision, duties of the supervisor and need for the operator to comply with the Licence.

Although the equipment is your property, there are two ways another amateur holding a UK amateur radio Licence may use it.

The **first way** is simply that you may lend it to them, so they operate it as if it were their own property, even if it remains in your house. When on air they must identify the station using their own callsign and obeying the terms of their Licence. They have responsibility for the station and may decide to add /A at the end of the callsign as they are transmitting from an alternative address but this is optional. If they were to leave your home and take the equipment away with them then operation is exactly as if it were their own property. For Licence purposes, irrespective of who owns the equipment or where it is operated from, even the actual owner's house, it is regarded as the station of the person who's callsign is given in identification.

The **second way** is operating under supervision. If you are supervising then you must be there with the operator and able to intervene if necessary. Your callsign is used so it is your station and you are responsible. The operator, must operate according to the terms of your Licence. That applies even if they hold a *superior* licence because they are using your callsign.

If, however, they are a Full Licensee you may be happy for them to supervise you. This means, that while they are monitoring what you are doing, you can operate the station using their callsign with all their privileges. In this instance they are taking responsibility for the station and will remain with you to ensure you do indeed obey their Licence terms.

In Summary

'Radio Equipment' is the equipment being used and identified by the operator's callsign.

Title of ownership is irrelevant; responsibility lies with the operator whose callsign is used.

The supervisor must be at the station in the presence of the supervisee to ensure that the terms of the supervisor's Licence are complied with. There is no objection to the supervisee sharing his personal callsign with the contact but there should be no confusion in identifying the station working under the supervisor's callsign.

Nets. You will have become aware that an amateur radio net is where a group of amateurs meet on a specific frequency to enjoy QSOs with each other. You or your supervisee are free to participate in nets once an initial contact has been made with a member of the group.

1C1 International Disasters, Non-amateurs Using Amateur Bands and Nets

Recall that in an International disaster messages may be passed, internationally, on behalf of non-licensed persons.

Recall that non-amateur stations involved in international disaster communications may also be heard on amateur frequencies.

International Disasters. During these events you may pass international messages on behalf of non-Licenced persons. You also need to be aware that non-amateur stations may be heard on the amateur frequencies. These people are usually assisting the recovery effort and you should not transmit on their frequency.

1C2 User Services, and Confidential Information in Messages

Recall that the Licensee may pass messages on behalf of a User Service and may permit them to use the Radio Equipment to send messages.

You are allowed to assist the User Services. At their request, you may transmit any information they provide unchanged that has been obscured to protect patient confidentiality. Similarly, you may allow a member of the user services to operate your equipment to send messages. The User Services are the British Red Cross, St John Ambulance, the St Andrew's Ambulance Association, the Women's Royal Voluntary Service (known as Royal Voluntary Service since 2013), the Salvation Army, and any Government Department. The identity of the User Services is not examined.

Raynet-UK (www.raynet-uk.net) is an organisation specifically set up to assist in the passing of messages at major events and emergencies. Raynet is not one of the User Services.

1D1 Causing Interference and its Consequences

Recall that transmissions from the station must not cause Undue Interference to other radio users.

Recall that the Licensee must reduce any emissions causing interference, to the satisfaction of a person authorised by Ofcom.

Understand that this may include a reduction in transmit power or any other action required to reduce emissions to an acceptable level.

Ofcom may require you to modify or restrict the use of your radio equipment either temporarily or permanently with immediate effect,
- If you breach the Licence conditions
- If you cause or contribute undue interference to other authorised radio equipment
- In the event of a local or national emergency
- If an investigation shows your equipment is the cause then you may be required to change your operating; perhaps to avoid a particular band or reduce power. You may also be required to stop operating temporarily. This might also occur if there is a national or local emergency even if your equipment is working properly.

You should keep your licence available for inspection.

1D2 The Importance of a well-kept Log.

Recall the occasions for mandatory log keeping.

Understand circumstances in which modification or cessation of operating of the station may be required.

Understand circumstances in which modification of transmitting equipment may be required.

You are no longer required to mandatorily keep a log but it is very much in your interest to do so for reference. It will help your advancement into the hobby and you will have a permanent record should there be interference problems with neighbours. Of course, there has to be a further but. Should Ofcom be involved in the investigation of complaints or otherwise you may be directed to keep a log and the format of the log such as mode, frequency, time, length of QSO, and power will need to conform to their specific request.

1E1 Remote Operation of the Equipment

Recall that the licensee may use any communication link for the purposes of Remote Control of their station and must ensure that:

- **any links used for the Remote Control of the Radio Equipment must be adequately secure so as to ensure that no other person is able to control the Radio Equipment;**
- **Remote Control links using Amateur radio frequencies must use frequency bands above 30MHz and must not be encrypted;**
- **transmissions from the Radio Equipment can be terminated promptly; and**
- **the Licence Number must be displayed on or next to any unattended Radio Equipment located other than at the address shown in the licence.**

Remote operation is an invaluable asset. When the hobby started it was the amateur radio operators who were considered to be the villains producing interference. Today the situation is reversed. Modern houses contain substantial amounts of electronic equipment which are constantly emitting radio waves across all the bands and giving amateurs an unpleasant background noise. By siting your transmitter somewhere free from all these distracting interferences you can remotely control it and reduce the irritating noise floor. There are important points to abide by,

- The remote control link is for your personal use only or somebody under your direct supervision.
- If the link fails the system must not be left transmitting.
- The link must be secure to prevent others accessing it but not encrypted.
- Any communication link (e.g. internet) may be used to control the remote Radio Equipment.
- If an amateur band is used it must be above 30MHz.
- Transmissions from the remote Radio Equipment can be terminated promptly.
- Your licence number must be displayed next to the remote Radio Equipment. Only Ofcom can use this to identify you.

You will have probably heard of Packet Radio. This is an example of a digital communications facility which is declining in popularity. It may be left running while the operator is away from the equipment. It allows for an automated device to receive text or data messages addressed to it by callsign, to acknowledge safe receipt and store those messages locally awaiting the return of the amateur owner. Acknowledging receipt involves transmitting while the equipment is unattended so needs to be covered by the Licence. Your Intermediate Licence does cover this type of unattended use but it is *not* examined.

1F1 Other Administrations and Recognition of your Licence

Recall that other Administrations (foreign countries) do not routinely recognise the Intermediate Licence.

Remember that other Administrations (foreign countries) do not routinely recognise the Intermediate Licence. Once again, there is a small but. It is known that some countries will allow radio operation to those who ask personally. That permission is effectively a temporary Licence issued by that country and is wholly outside UK jurisdiction. Should such permission be granted the operator will be required to obey the terms and conditions of that country.

1G1 Electromagnetic Fields

Recall:

- **that the average and peak transmit power level at which the EMF restrictions apply;**
- **when there is a need to reassess EMF compliance.**

At Foundation you covered the purpose of the EMF requirements which was to limit human exposure to RF signals to the recommended levels. These apply if your transmitted power level exceeds either 10W EIRP (see chapter 12) averaged over any 6-minute period or 100W EIRP peak, an instantaneous value even for a very short time. You may like to re-read the Foundation material on EMF, electromagnetic fields.

You will know that FM means you are radiating full power all the time you are actually transmitting. SSB only produces RF when you speak, and the power level varies. Other modes, such as CW, also have an average value depending on the proportion of the time the key is down sending a dot or dash. Spending more time receiving than transmitting will also keep the average down.

The body can absorb RF energy and warm up as a result. By averaging the power over six minutes the overall warming effect can be estimated and kept well below the level that the body's own temperature regulation system can comfortably handle.

It is also possible that a peak level of RF will cause a momentary sensation which may be distracting. To avoid that there is a limit on the peak level at 100W EIRP.

No doubt you performed an assessment when you set up your Foundation station. Even a dipole has some gain off the sides so 25W to a dipole can produce almost 41W EIRP. When you choose to transmit at different power levels you must re-assess your compliance distances and exclusion zones as at higher power levels these will have increased. Any future change should also be reassessed. If you conclude there is no change then all you need to do is record the change and the conclusion and keep it with the earlier assessment so it is always fully up to date.

1H1 Licence Schedule

Identify relevant information in Schedule 1 to the Intermediate licence.

A copy of the relevant part of Schedule 1 will be available during the examination. At Intermediate level you follow Table B of your licence.

Most bands allow 100W PEP (Peak Envelope Power) that is the power in the largest RF cycle transmitted, a voice peak on SSB but the constant value for FM. Don't forget however to check the schedule; It is 1W (0dBW) in the 135.7kHz band and 32W (15dBW) in the upper part of the 1.8MHz band. Some other bands have power limits below 100W so do check carefully when reading exam questions. Remember also that the airborne limit is 500mW EIRP (not ERP) and only allowed in Primary amateur bands as shown in the Schedule, Table B.

Not all bands are amateur Primary, some are Secondary and we must give way to the Primary user. Something else to check!

THE FULL LICENCE MANUAL FOR RADIO AMATEURS

By Alan Betts, G0HIQ

This book is the third course-book in the RSGB series for those interested in obtaining an amateur radio licence. In line with the progressive three-tier UK licence structure (syllabus 1.6) *The Full Licence Manual* completes the natural progression from Intermediate and Foundation Licence Manuals.

Fully revised to reflect all the recent syllabus changes, *The Full Licence Manual* contains all of the information required to move to the final stage of UK amateur radio licensing. This book is broken down into logical sections each written to match the Full licence syllabus. Licence conditions are covered in detail as are operating techniques and amateur radio safety. As you would expect, there are sections covering technical matters such as circuits, semi-conductors and more. The Transmitter and Receiver are covered in detail along with the material required for understanding the Software Defined Radio section of the syllabus. Feeders, Antennas and Propagation all get chapters of their own, as do Electromagnetic Compatibility and Measurements, All this means that *The Full Licence Manual* is the ideal companion to a formal training course. The book is also a useful reference source and many amateurs will find themselves referring to it long after they have passed their examination.

The Full Licence Manual is for everyone progressing to the Full licence from The Intermediate Licence and is the best route to success in the examination.

Size: 210x297 mm, 104 pages
ISBN 9781 9139 9558 4
Price £15.99

Also available
amazon kindle

Don't forget RSGB Members always get a discount
Radio Society of Great Britain www.rsgbshop.org
3 Abbey Court, Priory Business Park, Bedford, MK44 3WH. Tel: 01234 832 700

2: Operating Techniques

If you have been listening on the amateur bands, or have been active on air with your Foundation Licence, you will know that the bands are shared worldwide. You may have noticed at particular times, especially during contests, the bands are very busy and finding space to operate yourself may be difficult. Over the years, radio amateurs have developed procedures to help avoid these problems.

Abbreviations and Q-Codes

Most of the abbreviations and Q-Codes you will hear on the amateur bands date back to the time when CW was the only mode. By adopting a standard set of abbreviations, longer messages or commonly repeated phrases could be sent much more effectively.

Some of the Q-codes in common use are listed below, and you are required to know their meanings. Q-codes may be used as either a statement or a question.

- QRM = Man-made interference
- QRN = Interference from natural sources (e.g., lightning)
- QRO = High power
- QRP = Low power
- QRT = Closing down the station
- QSB = Fading
- QSL = Contact confirmation
- QSO = A contact with a station
- QSY = Change frequency
- QTH = Home location / location of the station

There are also some other abbreviations you may use or hear on air and are useful to know:

- DX = Long distance
- CQ = General call for a contact
- DE = From (often sent as the end to an over – 2E1ZZZ de 2E1CZR)
- R or RR = Roger (all received)
- SIG = Signals
- 5NN = 599 (a faster way to send a report of 599 in CW)

The Intermediate Licensee and Band Plans

You will recall from the chapter on Licence Conditions, that Schedule 1 of your Licence includes Table B in which are the legal requirements of an Intermediate Licensee.

1. Which part of the radio spectrum you may use.
2. Whether you have primary or secondary use.
3. The Status in UK to the Amateur Satellite Service
4. Maximum Peak Envelope Power you may use.

When you plan your CQ call you must ensure that you work within the above limits. There is, however, a further consideration. To use the bands efficiently we need to have a little more control over how they are populated and we do this with Band Plans. They divide each allocated amateur band into sections, and show what modes are acceptable in each section. The band plans are produced by the IARU (International Amateur Radio Union) after consultation with their member countries and associated radio organisations; each country may then alter them to match its own requirements. In some countries parts of their licence schedule are reserved to different classes of licence and observing the Band Plan is a licence condition.

Copies of the UK band plans are available from the RSGB website. The RSGB Band Plans are normally published annually in the February edition of RadCom and additionally reviewed mid-year. At the Intermediate level questions on Band Plans will be confined to the 2m and 20m band. The full 2m Band Plan can be found on the RSGB website www.rsgb.org and an explanation of its use can also be found there

A synopsis of the 2m Band Plan is given in Table 2.1. You are not expected to know it but it is essential that you understand how it should be used.

In the exam you will be given a copy of,

Reference Data for use in the Intermediate Level Examination

This should be downloaded from the RSGB website www.rsgb.org as document **EX308**

Formulae relating to the intermediate Syllabus are given in document EX308 and they are reproduced in Appendix 4.

You are not expected to know the details of the Intermediate Licence Parameters and the Band Plans in current use. The document contains representative information for use in the examination. Pre-reading is strongly recommended.

For your exam you must know,
- no SSB operation should take place in the 10MHz (30m) band.
- no contests shall be organised in the 5MHz (60m), 10MHz (30m), 18MHz (17m) and 24MHz (12m), bands.
- transmissions on satellite frequencies should be avoided for terrestrial contacts..

and be aware that questions on beacon and satellite frequencies will be limited to the 14MHz (20m) and 144MHz (2m) bands

Satellites

As well as the point to point radio communication you will probably have already used, and repeaters on the VHF and above bands, there are amateur satellites orbiting over 250km above the earth. Some of these satellites act similarly to repeaters, taking your signal and retransmitting it to a wider area. This is possible because the satellite is up so high, the area your signal will reach will be much larger. With some practice and the right

Picture 2.1: Amateur radio Satellite

Frequency Range MHz	Suggested Modes of transmission
144.0000 – 144.0250MHz	All modes (CW/SSB/MGM)
144.0250 – 144.1500MHz	Telegraphy (CW) and MGM
144.1500 – 144.4000MHz	Narrowband modes (CW/SSB/MGM)
144.4000 – 144.5000MHz	Propagation Beacons
144.5000 – 144.7940MHz	All Modes
144.7940 – 144.9900MHz	Machine Generated Modes (MGM) & Digital Comms
144.9900 – 145.1935MHz	Repeater Input Channels (FM/DV)
145.2000 – 145.5935MHz	Simplex Channels (FM/DV)
145.5935 – 145.7935MHz	Repeater Output Channels (FM/DV)
145.8060 – 146.000MHz	Amateur Satellite Service – All Modes

Table 2.1 Transmission Modes on the 2m Band

satellite, you can even make contacts between different continents.

Most amateur satellites are not stationary above the Earth's surface: they orbit around the Earth and are only above your horizon for a short duration at a time. You can only use the satellite when it is above your horizon: when it's below the horizon, the Earth itself will block your signals. Any stations you contact via the satellite will also have the satellite above their horizon. You can use satellite prediction software to see which areas of the earth can see each satellite at a certain time.

Because the satellite is moving towards or away from you, there is some frequency variation on the received signal known as **Doppler shift**. You will need to allow for this when selecting your operating frequencies.

To avoid too much interference from stations not intending to use the satellites, sections of some bands are set aside in the band-plans specifically for satellite use. You shouldn't use these frequencies unless you're trying to listen to or make a contact through a satellite.

The earth-to-space (uplink) and space-to-earth (downlink) frequencies will always be in different amateur bands, quite often in the 144MHz (2m) and 433MHz (70cm) bands. There are also some satellites that operate at least partially in the HF bands, for example, a 145MHz up-link with reception of the down-link on 29MHz.

You should remember that if you decide to transmit through these satellites your station *must* be able to receive on the frequency you transmit on as well as the down-link frequency.

Satellites are powered by solar panels which convert the energy from the sun's light into electrical energy to recharge the onboard batteries, so they only have a limited amount of power available.

Some satellites have very limited power, while others are very easily overloaded by strong uplink signals. This can cause interference to other users, or in some cases cause the satellite to switch off completely. As a rule of thumb, your down-link signal should be no stronger than the satellite beacon signal.

Digital Interfaces

There are many new and interesting modes that offer superb performance, allowing contacts when propagation conditions are poor or allowing images and audio to be transmitted on narrow HF bands. These, along with more traditional modes like CW and RTTY, can be transmitted and received using computer software and some type of digital interface to connect the computer to your radio.

Modern transceivers now often have a single USB port that interfaces the transceiver to a computer, allowing computer control of the radio and making the radio appear as a soundcard to the digital mode software.

If your radio doesn't offer this, then you can purchase or construct a simple interface. It will take a signal from your computer to switch the radio between transmit and receive, and also take audio signals to and from your computer soundcard to the radio.

Be aware that all the sounds produced by the computer will be sent to the radio, including various beeps, start-up sounds and music as well as noise generated by any other program that may be running. You should take great care to ensure these are not transmitted by your radio.

It is important that you correctly set the output levels from your computer sound card. If the level is too high, you will overdrive the radio leading to a worse (harder to decode) signal with lots of distortion. (Technically speaking, the sound card contains a Digital to Analogue (audio)

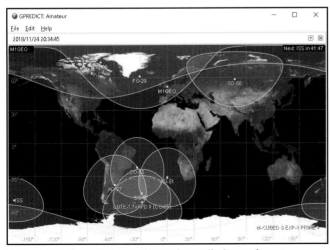

Picture 2.2: Gpredict satellite path prediction software

Picture 2.3: Tigertronics SignaLink USB digital mode interface

2: Operating Techniques

Converter, DAC, and its interface with the transmitter requires careful adjustment). There is also the potential of interference to other stations on adjacent frequencies. On most radios, you can use the ALC meter to check the incoming audio level from the computer sound card is correct – your radio manual should explain how to do this.

Contesting

It would be wrong to think that amateur radio is a hobby just for those who are very technically minded. There is a much lighter side to the hobby known as contesting.

Contesting usually involves a competition to see who can contact the most stations in a set period of time on certain bands and within specific rules.

Like all competitive activities there are rules and regulations which you need to be aware of, The RSGB website and Yearbook will provide you with lots of background information to get you started. More often than not your club will have a keen group of contesters.

For your exam you must know,
Some of the common international call sign prefixes,

- *Eire*   EI
- *France*   F
- *Italy*   I
- *Japan*   JA
- *Netherlands*   PA
- *Canada*   VE
- *Australia*   VK
- *USA*   W
- *New Zealand*   ZL

There are awards available for achievement in working (making contact with),
- *continents*
- *countries*
- *islands*
- *prefixes*
- *locator squares*

including variation in;
- *frequency*
- *bands*
- *power*
- *mode*

A contact will involve the exchange of information,
- *signal report*
- *serial number*
- *location*

All these points, when properly logged, help you to progress in your hobby but more importantly will validate your contact.

3: Tools, Construction & Safe Practice

Tools, Construction and Safe Practice.

The safety and wellbeing of yourself and others around you should always be your primary concern. It is impossible to highlight every potential risk and as you are responsible for your actions your planning must include an examination of what could go wrong. This will inform you as to what you need to do to work safely. Technically, this foresight is known as a risk assessment.

The most important piece of equipment will be a pair of good quality safety glasses. Any activity involving cutting and shaping will involve swarf and dust which can cause irreparable damage to your eyes. It would be unwise to think that prescription glasses can afford the necessary protection. Light weight glasses use prescription lenses made of plastic which are easily marked and very expensive to replace. Goggles can be worn over prescription glasses.

Soldering

A circuit contains many components and it is essential that there are good electrical joints connecting them. This is achieved by using a hot soldering iron to melt a thin rod of solder so that it fills the gap at the point of connection with molten solder. **Picture 3.1** shows a component being soldered onto a printed circuit board (PCB)

If your technique is good the molten solder cools, solidifies and a good connection is made. This happens when the molten solder flows freely "wetting" the metals and cooling to a clean shiny surface as shown in **Picture 3.2.**

Always examine your joints carefully. Sometimes the solder does not flow properly and a good electrical bond is not made. This is known as a *dry joint*.

A soldering iron is generally heated electrically. The tip will reach temperatures of 300 to 400°C. For most purposes a 15W soldering iron is adequate. When not in use it should supported in a stable stand as shown in **Picture 3.3**

Care of the tip of the soldering iron is important. As it is hot and in contact with the oxygen in the air it will become dirty and covered in a layer of oxide. Keep it clean and "wet" it with a small dab of solder. The molten solder on the tip will help to give good heat transfer to the items being soldered. Similarly, it can be advantageous to tin the wire of the components being soldered together. This will also help the flow of heat and molten solder.

Solder is a mixture of metals, most commonly tin and copper or tin and lead. Solders not containing lead, a poisonous metal, comply with modern regulations. Lead free solder compositions contain silver but are not so easy to work with. **Picture 3.4** Shows a cross section of cored solders.

The core contains a flux. This is required to remove any oxide which may be present on the metal surface preventing the formation of a good joint. Amateurs commonly used "rosin cored solder". Rosin is obtained from pines and plants and can be used as a flux. Fumes from the molten flux can be very irritating and it is always a good idea to work in a well-ventilated area to prevent problems with eyes and breathing.

Good solder contacts are made between tin, copper and brass. Fortunately, most radio work involves soldering components onto printed circuit boards where the connecting tracks are made of copper. However, aluminium and stainless steel are difficult to solder and special techniques are used.

Further information associated with the risks of fumes from solder can be obtained from the Health and Safety Executive website:
www.hse.gov.uk

- Solder Fumes and You (INDG248)
- Controlling health risks from Rosin Based Solder Fluxes (INDG249F)

Picture 3.3: A stable stand is essential to support the soldering iron when it is not in use

Picture 3.4: A cross section of multicore solders showing channels where flux is stored

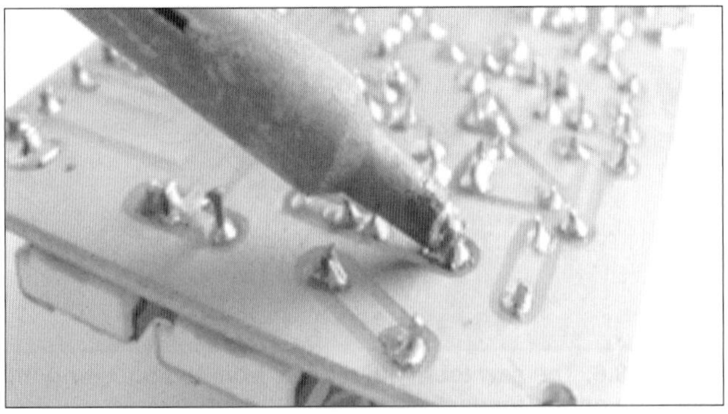

Picture 3.1: Warming the joint and letting the solder flow freely

Picture 3.2: Good quality soldered joints are bright and shiny

- Soldering iron & stand
- Solder sucker and/or solder braid
- Side cutters
- Wire stripper
- Sharp modelling knife
- Flat blade screwdriver (2-3mm)
- Cross point screwdriver (2-3mm)
- Small pliers
- Multimeter
- A clear rule or tape measure
- A magnifying glass

Table 3.1 What a basic tool kit for constructing radio-related projects might contain

Tools

Hand Tools and Power Tools

Even the simplest of hand tools can cause injuries such as cuts and bruises, and power tools can be lethal if not correctly used. Being aware of this and taking a few sensible precautions will help you minimise any accidents or injuries.

All tools should be properly maintained and stored safely. Very old power tools may have metal bodies and will not be "double insulated". The importance of double insulation is explored in Chapter 8, Good Radio House Keeping. You are reminded to ensure that all power cables are in good condition.

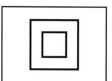

Fig 3.1: The double insulated symbol

Your toolkit

Most amateurs, at sometime or other, like to do some experimental work and this will necessitate a selection of tools. Invariably, tools are a personal choice. What feels good in your hands may not suit someone else. Seek advice, choose them carefully and buy the highest quality you can afford. Cheap tools may let you down after a lot of use.

If you are building cases you may need tools such as a hacksaw, drills and files.

You will also need a suitable workspace. Not everyone will have the use of a workshop, so if using a garden bench or coffee table you should use a craft mat to protect the surface.

You may also find community workshops and maker groups such as the various 'Hack Space' groups a good way to access a comprehensive workshop space.

Working with tools

Many tools have sharp points and edges, so you need to be careful when using them. Always pick your tools up by their handles and think about other people who may be near and may not be aware of risks. Remember that you can trip over power tool leads.

Knives, saws, files and drills have cutting edges which are much harder than flesh. They should always be used such that the cutting movement is in a direction away from the body and your fingers are always behind the blade. There is a temptation to support small pieces of work in one hand and the tool in the other hand. Should the tool slip you will almost certainly damage the hand holding the workpiece. This is particularly so when using a screwdriver.

Using a vice

Make sure that whatever you are machining is securely fixed. A vice can stop the workpiece slipping or spinning out of control. It can be permanently fixed on your workbench, or clamped onto your temporary workspace.

Drilling

Hand drills are less hazardous than power drills. When drilling, keep your body clear of any sharp tools. If using a power drill, remember to remove the chuck key once you have tightened the drill bit, or it could fly out when the machine is started.

Use a centre punch to make a small indentation where you intend to make the hole. The drill bit is far less likely to slip, making it safer and more likely the hole is drilled exactly where you want it. Drill a small pilot hole first, then use progressively larger drill bits until you have the size you want – this is not only safer but also gives a neater finish.

When using sheet metal, spiral shreds can spin off – these are known as swarf – and can be extremely sharp. Swarf can also eject at high speed from the drill bit, so wear safety goggles or spectacles. After drilling each hole, brush the swarf away using a paint brush, and once the drilling is complete use a vacuum cleaner or dust pan and brush to clean up. Never try to blow away the swarf as this could result in eye injuries.

Generally, gloves are *not* recommended when drilling: they make it harder

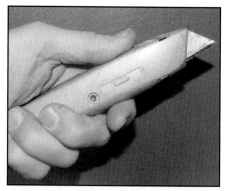

Picture 3.5: A firm hand grip behind the blade will prevent injury

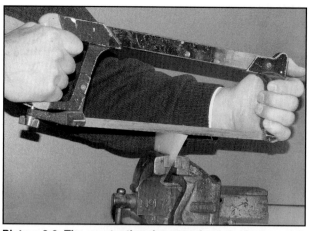

Picture 3.6: The saw teeth point away from the body and the saw cuts on the forward movement.

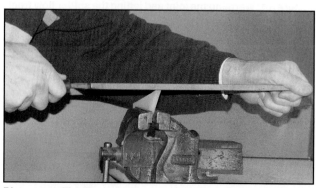

Picture 3.7: The file removes material in the forward movement away from the body.

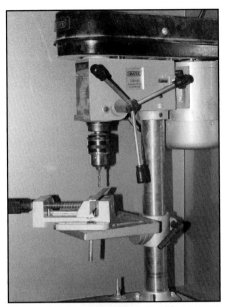

Picture 3.8: Using a pillar drill with a securable vice ensures that the workpiece cannot slip. A very important safety point is to ensure that the chuck key is removed from the chuck before the drill is used. Damage to one's self and property can result as the key is spun away from the machine. Do not forget that other power tools will have locking mechanisms to hold the tool securely in place and similar precautions are necessary.

Picture 3.9: Notice that both hands are used; one hand ensures that the blade does not slip in the screw slot and the other provides force to drive in the screw

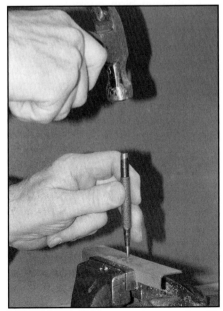

Picture 3.10: Using a centre punch with a vice

to grip tools, and if they get caught, they may drag your hand into the rotating drill! Always reduce the risk of movement of your workpiece by having it securely clamped or held in a vice.

A handy piece of equipment is a pillar drill. This may be an electric hand drill fixed to an upright stand and a handle is used to raise and lower the drill onto the work, or it may be a custom-made machine. The drill is fixed to the workbench for safety and enables your hands to be clear of the workpiece. This gives you more control over the operation.

Some thoughts on the construction of a transceiver

A good circuit design is only the first step on the road to building a successful transceiver. There are some very important practical considerations as to where you position components on the chassis.

Electronic components and the sub-assemblies from which they are made can have associated electrostatic, magnetic and RF fields. The interaction of these fields with other components inside the transceiver can lead to annoying interference problems and the plan for where the different parts of a transceiver are placed is of great importance.

If you have looked inside a transceiver you will most probably have observed that some of the components have been sited in thin metal enclosures. Aluminium is most commonly used for this and its action can be likened to the shadows formed in sunlight. Any component on the opposite side of the screen will be in the shadow of unwanted fields that would cause undesirable coupling. The aluminium is connected to the metal chassis and any induced currents will go to earth.

Intermediate Frequency Transformers, as shown in **Picture 7.3** in the Transmitters & Receivers chapter are enclosed in a metal case. Power transformers generate a magnetic field and if they are not adequately screened there can be an irritating mains hum. Integrated circuits are often protected and in the older valve operated circuits it was not uncommon for them to have an aluminium sheath. Oscillator circuits are screened. The capacitive effect of a hand close by can cause the output frequency to deviate unacceptably.

It is always advisable to keep connecting wires between the different circuit units of a transceiver as short as possible and to use shielded cable. Further thought is necessary to prevent the ingress of coupling fields through holes or joining seams in the metal plate.

Screening is a two-way process. It is not surprising that a transceiver has a metal case. It will help to prevent radiation from within interacting with other external electronic equipment and to stop the RF radiating from your antenna interacting with the internal components.

Electrical

The Mains Supply

Being a UK licensed radio amateur allows you to build your own equipment. It is always advisable not to work on live apparatus but there may be occasions when it is unavoidable. There is always a serious risk of fire, perhaps an explosion, a shock and electrocution. An electric shock is not necessarily fatal but the muscular contractions that follow may cause you to fall or touch more live parts. Electric current flowing through the body can cause deep burns. There are some sensible precautions you can take,

- Think through what you intend to do; make a risk assessment and write down for others what your plan of work is.
- Use a Residual Current Device (see later)
- Use insulated tools and rubber gloves.
- Do not work alone; someone should be nearby if aid is needed.
- Ensure that your shack has a master switch. In the event of an accident rendering you unconscious anyone coming to your assistance can cut off the power and will be confident they have no additional risk of electrocution.

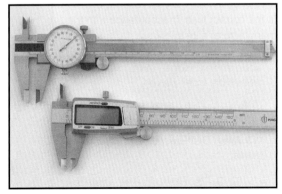

Picture 3.11: Vernier Calipers are very useful when making accurate measurements

Overhead power lines

High voltage overhead power cables carry electricity around the country to homes and businesses. Any contact with these cables is likely to result in fatal injuries. The high voltages can arc across several metres. Be extremely careful when using ladders or installing antennas or masts around overhead power cables. Make sure you check the area around where you are working.

Fuse rating

You have probably seen fuses before – a short glass or ceramic tube with a short thin metal strip inside. It is a safety device found in plugs and electrical equipment. The size of the metal strip is chosen so that it melts when a certain current flows through it, breaking the circuit and stopping the current.

It is extremely important that the correct fuse is fitted to electrical items in the home and workshop. If too small a fuse is used, it will blow in normal use; too high and it will not disconnect the equipment even if a dangerous fault exists.

Fuses are cheap and simple, but they are slow, since they require a piece of metal to get hot and melt.

Fuses provide good protection against too much current, but they are not enough to prevent a fatal electric shock. It only needs 0.05 Amps of current to stop a human heart. A 5 Amp fuse would provide no protection if a fault on your equipment only caused a current of 3 Amps to flow.

The main purpose of a fuse is to safeguard the equipment. By preventing excess current we can reduce the risk of overheating and, possibly, fire. This means we have to choose a fuse that will disconnect the circuit at a specified level of current. For domestic appliances, there are three common ratings for fuses; 3A, 5A, and 13A. Some examples of typical use are given below but you must remember they are not definitive; it is your responsibility to fit the correct fuse in accordance with the manufacturer's instructions.

3 Amp: *Clock radio, DVD player, fan, food mixer, fridge, lamp, PC, radio, television, video player*

5 Amp: *Coffee maker, hi-fi, microwave oven, toaster*

13 Amp: *Dish washer, hair dryer, iron, kettle*

Calculating the value of fuse required is necessary knowledge for your exam. Manufacturers will specify the power requirements of their product. If your instruction manual says that your transceiver consumes 500 Watts and we know that the mains supply is 230 Volts,

$$\text{Watts} = \text{Amps} \times \text{Volts}$$

$$500 = \text{Amps} \times 230$$

$$\text{Amps} = 500 / 230 = 2.17A$$

We fit the next higher rated fuse which is 3 Amp

Fast and Slow Blow Fuses

As well as the common ceramic type fuse we put into our mains plugs there are two other types that are important; those that blow quickly and those that blow slowly.

In equipment with delicate components such as integrated circuits it is necessary to use a fast blow fuse to break the circuit before the over current or spike causes damage

In circuits where there is a large surge in current during power on/off, for example electric motors, it is appropriate to use a slow blow fuse. This feature of drawing a high current on start-up is not considered to be a fault but part of normal circuit operation. These fuses are more expensive and sometimes called *time delay fuses*.

Fast blow and slow blow fuses are *not* interchangeable; they have different properties designed for specific applications. The most important point to remember is that the fuse used in any equipment must be that specified by the manufacturer.

Power Supply, Domestic & Field Day

This material is not examinable but it is designed to add value to your appreciation of the hobby. We need to start with an

Picture 3.12: Examples of 13, 5 and 3 Amp fuses used in plugs

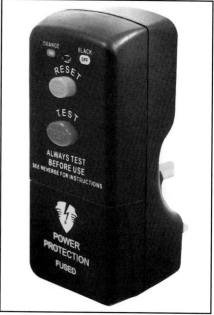
Picture 3.13: A portable RCD. Note the pins for insertion into the mains socket are at the back.

overview of the domestic mains supply. Regulations are always under review but the present expectations are,

The Mains supply, the cable from the road, leads to the supplier's sealed fuse and is followed by the consumer's meter.

After the meter is the Mains Switch and this leads into the Consumer Unit. Inside the Consumer Unit are several RCBOs for the separate circuits in your home. An RCBO is a Residual Current Circuit Breaker, with Overload protection. Each RCBO contains an MCB, Miniature Circuit Breaker, and an RCD, Residual Current Device.

a. The MCB gives protection against high current overload and short circuit. It contains an electromechanical device which releases a latch to open a switch thereby cutting the supply of large currents that can damage the equipment, cause rapid heating and possibly fire.

b. Should you accidentally touch a live wire the current flowing through your body will cause an imbalance in the live and neutral wires. The RCD is sensitive to current imbalances of about 30mA and will automatically break the circuit in about 25 to 40 milliseconds. It will increase your chances of surviving electrocution.

Do you have this level of protection when operating outdoors on event like Field

Day? The answer to this comes in your risk assessment. Many modern transceivers can be operated from heavy duty batteries so you may not need a mains supply. If you are going to use a Portable Generator you should look carefully at its specification. A High Quality portable generator for Field Days should have a built in RCBO or at the very least MCB protection or a reset fuse.

Whether you are working in the domestic or outside environment it makes very sound sense to ensure that your equipment has that extra level of safety given by a portable RCD. They are obtainable from most DIY stores.

Lightning

Thunderstorms are a threat to the radio amateur operator. Aside from the more commonly recognised hazard to life due to direct lightning strikes, static electricity can also build up on antennas and cables or be induced by nearby strikes.

Nothing will protect you or your shack from a direct strike, but gas discharge arrestors or spark gaps in antennas feeders can help with nearby strikes. These work by presenting an open circuit at normal transmitter powers, but break down and conduct when a high voltage is present (such as during lightning activity) briefly becoming a low resistance and dissipating the energy from the strike.

Static electricity build-up can be avoided by connecting a large value resistor (typically greater than 100 kilo-Ohms) across the antenna. This does not affect the antenna, but provides a path to ground for static electricity.

Electromagnetic Radiation

Concentrated RF energy can be very dangerous if not treated with respect. As you will remember from your Foundation level studies, you should never touch an antenna while transmitting due to the potentially high voltages and currents. RF currents can cause deep and serious burns, and RF voltage packs a nasty zap! Antennas on low power transmitters such as those found on a handheld radio are usually safe, but it is best to avoid touching any antenna while it is transmitting.

Too much RF energy can cause dangerous heating within the human body, in a similar way to how a microwave oven functions. An amateur radio station is unlikely to reach such a high level of radiated energy, but, you should be extremely careful if working with microwave radiation – never look into a waveguide or stand in front of a dish antenna while transmitting, since these focus the RF energy into an intense beam, further increasing the heating effects. Your eyes and your brain are particularly susceptible to such heating.

If you are at all concerned about electromagnetic radiation safety, you can find out more from the National Institute for Health Protection (NIHP) and the International Commission on Non-Ionising Radiation Protection (ICNIRP). You need to know the names of these bodies but you will not be asked questions about the levels themselves.

Transmitting within the domestic environment is strongly discouraged. Modern houses and gardens, all too often, do not have a great amount of space for erecting antennas. Consequently, there may be a temptation to mount the antenna in or very close to your shack. This will lead to you being exposed to RF fields greater than the agreed limits. You need to be able to assure yourself that you are working safely. Seek advice; talk to those with experience. Similarly, many amateurs have found the use of the loft space suitable for mounting an antenna but consideration is required for other people who may be in the house.

Component Symbols

Component	Unit	Symbol
Resistor, fixed	Ohm	
Resistor, potentiometer	Ohm	
Resistor, pre-set	Ohm	
Resistor, variable	Ohm	
Capacitor, fixed	Farad	
Capacitor, electrolytic (polarised)	Farad	
Capacitor, variable	Farad	
Inductor, fixed	Henry	
Inductor, with core	Henry	

Component	Unit	Symbol
Transformer	nil	
Quartz crystal	Hertz	
Semi-conductor diode	nil	
Diode, variable capacitance	nil	
Diode, light emitting	nil	
Transistor, bipolar NPN Note Transistors can be drawn with or without the circle	nil	
Transistor, field effect (FET) Note Transistors can be drawn with or without the circle	nil	
Earphone	nil	
Microphone	Ohm	

Component	Unit	Symbol
Loudspeaker	Ohm	
Cell	Volt	
Battery	Volt	
Bulb (lamp)	Watt	
Switch (SPST)	nil	
Switch (DPST)	nil	
Fuse	Amp	
Antenna	nil	
Earth	nil	
Ground chassis	nil	

3: Tools, Construction & Safe Practice

4: Basic Electronics

We will start this chapter with a quick re-cap of material you should remember from the Foundation Licence, and build upon it to prepare you for the Intermediate licence. You should recall the definitions of current, voltage, Ohms law and electrical power from your Foundation studies.

Current and Voltage

Current
Current is the flow of charged particles – normally electrons – past a given point each second. Current is measured in **Amperes**, or **Amps** for short and given the symbol **I** in calculations.

Current: DC and AC
There are some important differences between the movement of current in DC, AC and RF circuits. A short revision of DC and AC current will be advantageous before we look at current in an RF circuit.

DC Current
In a DC circuit the direction of the current and its value does not change (so long as the battery is not exhausted). This is shown in **Figure 4.1**

AC Current
Let us imagine that we have replaced the cell by a 50Hz AC supply.

As there are 50 complete cycles for every second. One cycle will take one fiftieth of a second which is 0.02 secs or 20ms.

In **Figure 4.2** we can see how the direction of the current changes with time.

Using **Figure 4.3** we can consolidate these ideas.

At point "A"
the current at any point in the circuit is 10 Amps and flowing in a clockwise direction

At point "B"
the current at any point in the circuit is 4 Amps and also flowing in a clockwise direction.

At point "C"
the current at any point in the circuit is 5 Amps but flowing in the anticlockwise direction.

At point "D"
the current at any point in the circuit is 8 Amps but also flowing in the anticlockwise direction

Voltage
The unit of electrical potential is the Volt. Specifically, it is the unit of potential difference (PD) between two points. They could be in a circuit, across a battery or between the ground and a storm cloud.

Cells and Batteries
In a cell a chemical change takes place to produce a potential difference. The value of the potential difference depends on the chemistry taking place. There are many different chemical reactions we can choose from and the majority of them will produce a potential difference of about 1.5 Volts.

Obviously 1.5 Volts may not be enough for our needs. We can overcome this problem by linking the cells together in series to give a battery. The total potential difference will be the sum of the individual potential differences as in **Figure 4.4**

Alternatively, we can connect the cells in parallel as in **Figure 4.5** When we do this the potential difference will not increase, it will still be 1.5 Volts, but the capacity of the battery will be greatly increased.

Electrical Power
Power is measured in **Watts (W)**, and in electric circuits the power can be calculat-

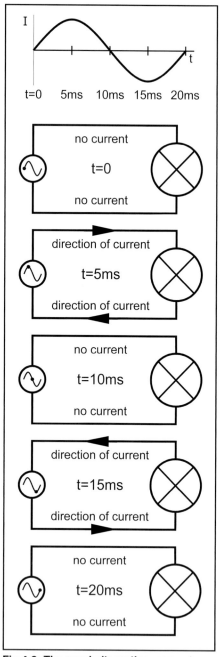

Fig 4.2: The word *alternating* means to go back and forth or to swing between two states or conditions. The arrows show the changes in direction of the current

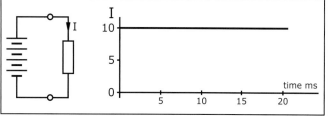

Fig 4.1: A graph showing how the current in a DC circuit remains constant

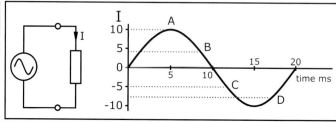

Fig 4.3: A graph showing the change in current with time for a 50 Hz supply

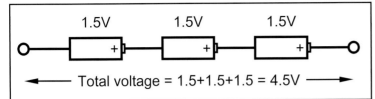

Fig 4.4: Cells connected in series

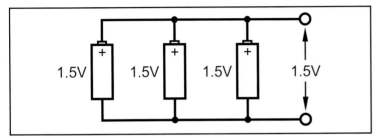

Fig 4.5: Cells connected in parallel

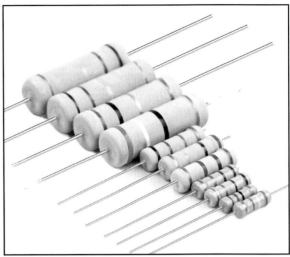

Picture 4.1: Fixed value resistors

ed by multiplying the current and voltage.

Power (Watts) = Current (Amps) x Potential Difference (Volts)

You should be able to re-arrange the formula to calculate any of the values if you are given the other two.

Example
A circuit draws 2 Amps of current at a voltage of 17 Volts. What power does the circuit use?

$$P = I \times V = 2 \times 17 = 34 \text{ Watts}$$

There is also another way with which we can calculate power. Ohm's Law, the next topic, tells us the relationship between Current, Voltage and resistance,

Current (Amps) = $\frac{Volts}{Resistance}$

or, more commonly, $I = \frac{V}{R}$

We said, above, P = I × V
and by substitution we now have

$$P = \frac{V}{R} \times V \quad \therefore \quad P = \frac{V^2}{R}$$

You will appreciate later that this is very important to radio engineers. Whenever the voltage is doubled the power goes up by 2^2. A doubling of the voltage gives us a fourfold increase in power

Resistors and Resistance

Resistance is opposition to current flow. For a given potential difference, a larger resistance will allow only a small current to flow through it, while a low resistance would allow a larger current to flow. Resistance is measured in **Ohms** and is given the symbol Ω. One Ohm is a small resistance and you will more likely encounter resistances of hundreds, thousands or millions of Ohms. Thousands of Ohms are given the prefix 'kilo' and millions of Ohms are given the prefix 'mega'.

You will see reference to 10 kilo-Ohm resistors or 3.3 mega-Ohm resistors. These may be shorted to 10k or 3.3M. Sometimes the letter will replace the decimal point, e.g. 3.3M may be written 3M3, or 4.7k may be written 4k7.

When building circuits we need to identify and select resistors of the correct resistance. You need to know that there is a code of colours that enables us to do this.

Ohm's Law

Ohm's Law defines the resistance of an object as the potential difference across the object divided by the current though it.

R (Ohms) = PD (Volts) / I (Amps)

$$R = \frac{V}{I}$$

Example
A circuit allows a current of 0.1 Amps to flow though it with a potential difference of 10 Volts across it. What is its resistance?

$$R = \frac{V}{I} = \frac{10}{0.1} = 100 \Omega$$

As with power, you should be able to calculate any value, using the triangle in the **Figure 4.7**

Resistors in Series and Parallel

If two resistors are connected end to end, they are said to be in **series** and their total resistance is the sum of the two resistors.

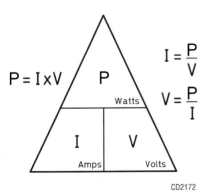

Figure 4.6: Cover the letter of what you want to know and the remaining letters show you how to arrange the formula

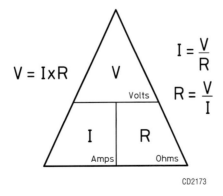

Fig 4.7: Cover the letter of what you want to know and the remaining letters show you how to arrange the formula

4: Basic Electronics

Example

A 1000 Ohm resistor is connected in series with a 330 Ohm resistor. What is the total resistance?

$$R = 1000 + 330 = 1330\ Ohm\ or\ 1.33k\ Ohm$$

Resistors in parallel share the current depending on their individual resistances. For resistors in parallel, the total resistance will always be lower than the value of the smallest resistance.

The formula for some number (*n*) of resistors in parallel is

$$\frac{1}{R_{total}} = \frac{1}{R_1} + \frac{1}{R_2} + \cdots \frac{1}{R_n}$$

Example

A 500 Ohm resistor is connected in parallel with a 1000 Ohm resistor. What is the total resistance?

$$\frac{1}{R_{total}} = \frac{1}{500} + \frac{1}{1000}$$

$$\frac{1}{R_{total}} = 0.002 + 0.001 = 0.003$$

$$R_{total} = \frac{1}{0.003} = 333.333\ Ohm$$

Resistor Values

A resistor is generally made from a short length of carbon enclosed in a ceramic tube. Two stiff wires, for connection, are attached to each end of the carbon rod.

The ceramic body of the resistor varies in length but is generally not longer than about half an inch. Around the body are coloured rings and you will see that they are closer to one end than the other. Be careful to remember that resistors

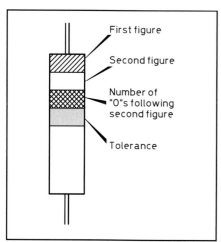

Fig 4.8: Resistor layout

with four and five bands are used. The colours denote numerical values as can be seen in **Table 4.1**

You will be given the colour code in the exam but it will be to your advantage to be familiar with the colours and their values. Remember that we always start with the colour nearest the end of the resistor.

The first two bands give us the first part of the resistor value. For example if they were brown and green the value starts with the number 1 and is followed by 5. We would then look at the colour of the third band as this indicates the number of 0s that need to be added. If it is orange there are three 0s to follow and the resistance would be 15,000Ω or 15kΩ.

A resistor with the colour banding, yellow, violet, red denotes the numbers 4 and 7 followed by two 0s ; 4,700Ω or 4.7kΩ.

Some special points about the third band.

Apart from telling us the number of 0s to add it is really telling us the power of ten to which the first two significant Figures must be multiplied. Using the colour code we can see

Red, tells us to multiply by 10^2 or 100 and add two 0s

Orange, tells us to multiply 10^3 or 1000 and add three 0s

Green, tells us to multiply by 10^5 or 100,000 and add five 0s

There are three colours, two of which we have not met before, that can cause difficulty,

Black, indicates that there are no 0s to follow the first two numbers. A resistor with the colours yellow, violet, black has a value of 47Ω.

Gold, indicates that the multiplier is 10^{-1} or 0.1

If the colours are red, violet, gold we have 2 followed by 7 or 27 which we multiply by 0.1 to give 2.7Ω

Silver, indicates that the multiplier is 10^{-2} or 0.01

If the colours are brown, black, silver we have 1 followed by 0 or 10 which we multiply by 0.01 to give 0.1Ω

The colour of the **last ring** is very important because it is an indication of how much the real (or experimentally determined) value of the resistor could vary from its stated value. This colour gives us the tolerance of the resistor. There are four

Colour	Value
Black	0
Brown	1
Red	2
Orange	3
Yellow	4
Green	5
Blue	6
Violet	7
Grey	8
White	9

Table 4.1: Resistor colour codes

recognised tolerances,

Brown 1%
Red 2%
Gold 5%
Silver 10%

For the exam you need to know the importance of the Gold and Silver bands.

Examples

Let us imagine a resistor has the colours, brown, green, orange, brown.

It will have a value of 15kΩ ± 1%.

1% of 15,000 is 150 and the actual value of the resistor will be between
15,000 − 150 = 14,850Ω
and 15,000 + 150 = 15,150Ω

Similarly, a resistor with the colours brown, green, orange, silver will have a value of 15kΩ ± 10%

10% of 15,000 is 1,500 and the actual value of the resistor will be between
15,000 − 1500 = 13,500Ω
and 15,000 + 1500 = 16,500Ω

Be careful of the importance of gold and silver bands. If they are the third band they are multipliers but if they are the fourth band (or fifth; see later) they indicate tolerance.

In many cases the choice of resistor tolerance will depend on the design of your circuit and may not be too critical. However, components manufactured to close tolerances can become more expensive.

When more precision is required we use an additional coloured band. Instead of working with four bands we work with five bands; the first three bands give us a three Figure number, the fourth band tells us the number of 0s and the last band gives us the tolerance.

In the first example we looked at the colour banding of a 4.7kΩ resistor. It might be that we wanted to use a 4.75kΩ or 4,750Ω resistor. What colour banding would we be looking for? The first three colour bands would be yellow, violet and green and because we want one 0 the last band would be brown. The colour bands would be yellow, violet, green, brown

Example
As another example. A resistor with the colours, green, yellow, blue, yellow, gold.
will have a value of, 5,460,000Ω
with a tolerance of 5%
or more simply 5.46MΩ ± 5%

If you are working on home construction projects you may come across resistors with a sixth coloured band. This is certainly not required knowledge for the exam but the extra band tells us how the value of the resistor changes with temperature.

In theory there should be a resistor of any value you require to fit your circuit design perfectly. It is impossible for a manufacturer to produce such a great range of different values and the electrical industry has agreed to produce a range of resistors with fixed values which are acceptable for most purposes. (At the Intermediate level this is restricted to what is known as the E12 series and you are not required to know any of the values).

Voltage Divider
It is possible when designing circuits that the power supply gives us a much higher voltage than we need. We can reduce the higher voltage using resistors arranged in what is commonly termed a *voltage or potential divider*.

It is usual to denote the voltage applied to the potential divider as V_{in} and the reduced voltage as V_{out}. In **Figure 4.9** there are ten 1Ω resistors connected in series. V_{out} is changed by moving the tapping point to any of the connections between the resistors.

We can see that the eight resistors above the tapping point have a total resistance of 8Ω and the two resistors below the tapping point have a total resistance of 2Ω and the total resistance is 8 + 2 = 10Ω

V_{in} is the voltage across all 10 resistors, which means the voltage across each individual resistor must be
$\frac{Vin}{10}$ volts

Therefore, in the diagram, where the pointer contacts the wire where the resistance is 2Ω.

V_{out} is $2 \times \frac{Vin}{10}$ volts

and this can be re-written as

$V_{out} = V_{in} \times \frac{2}{10}$ volts

- The eight 1Ω resistances in series have a total resistance of 8Ω and can be thought of as being single resistor of value, R1
- The two 1Ω resistances in series have a total resistance of 2Ω and can be thought of as a single resistor of value, R2
- R1 + R2 = 10Ω

and we now have a general formula

$V_{out} = V_{in} \times \frac{R2}{R1+R2}$

We can demonstrate the use of the formula with three examples.

Example 1
What is V_{out} given that V_{in} is 50Volts, R1 is 10Ω and R2 is 5Ω?

$V_{out} = V_{in} \times \frac{R2}{R1+R2}$

$= 50 \times \frac{5}{5+10} = 50 \times \frac{5}{15}$

$= 50 \times \frac{1}{3} = 16.6 \text{volts}$

Example 2
What is V_{in} given that V_{out} is 20 Volts, R1 is 10Ω and R2 is 30Ω ?

$V_{out} = V_{in} \times \frac{R2}{R1+R2}$

$20 = V_{in} \times \frac{30}{10+30}$

$20 = V_{in} \times \frac{30}{40} = V_{in} \times 0.75$

Hence, $V_{in} = \frac{20}{0.75} = 26.66$ volts

Example 3
V_{out} is 8 Volts, V_{in} is 50 Volts and R2 is 10Ω. What is the value of R1?

$V_{out} = V_{in} \times \frac{R2}{R1+R2}$

$8 = 50 \times \frac{10}{10+R1}$

$80 + 8 R1 = 500$

$8 R1 = 420$

$R1 = \frac{420}{8} = 52.5 Ω$

Potentiometers
Choosing the values of R1 and R2 may be difficult because components are not always manufactured to the exact values that we want. Fortunately, we have a much more convenient form of voltage divider called a *potentiometer*. **Pictures 4.2** and **4.3** show the external and internal views

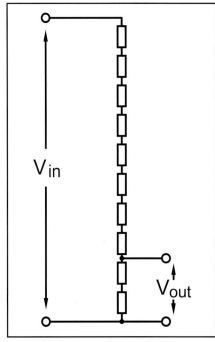

Fig 4.9: A simple potential Divider made from 1Ω resistors

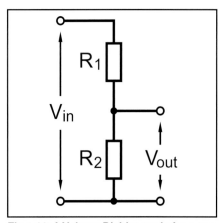

Fig 4.10: A Voltage Divider made from two fixed value resistors

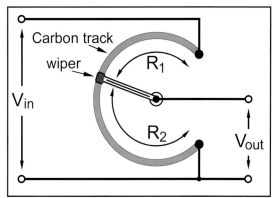

Fig 4.11: The internal mechanism of a potentiometer

of a potentiometer. Its construction can be shown as in **Figure 4.11**.

As the wiper is rotated the values of R1 and R2 change, the value of R1 + R2 remains constant, and V_{out} can be adjusted to the required value.

Potentiometers are generally linear. The change in resistance is directly proportional to how far we advance the wiper on the carbon track. For volume controls in audio equipment the resistance would vary in a logarithmic fashion. This is because the human ear responds to loudness of sound logarithmically.

Potential Difference, Electromotive Force and Source Impedance

A battery only holds a limited amount of electrical energy; the electrical energy is produced by a chemical reaction inside the battery, and as the chemicals are used up the available electrical energy falls.

When choosing a cell it is an advantage to have a measure of how much energy it can store. This is very easily expressed in **ampere hours** (Ah). One ampere hour is when a current of 1 Amp flows for 1 hour. A car battery has a capacity of 50 ampere hours. This means that it could supply a current of 50 Amps for one hour or 5 Amps for 10 hours. By comparison an AA cell is rated at 2 to 3 ampere hours.

A vehicle battery is designed to store a very large quantity of energy. It is a very

Picture 4.2: A continuously variable resistor or potentiometer.

Picture 4.3: Inside a variable resistor or potentiometer. As the spindle is rotated a contact moves against a resistive layer of material. This type of variable resistor is commonly found in volume controls in audio amplifiers.

popular power source amongst amateurs but its potential dangers need to be recognised. Carelessness can lead to very high currents being drawn causing excessive heating in wires and possibly fire. The battery electrolyte is strong sulphuric acid which is very corrosive. Any spillage should be neutralised and attended to by a responsible person. These batteries should always be recharged in a well ventilated area. Hydrogen gas is produced during charging and this forms an explosive mixture with air.

The rate at which electrical energy can be taken from a battery is limited by the construction and the chemistry of the battery. A car battery can deliver a much greater current than a 9 Volt PP3 battery for example.

You can think of a real battery as a perfect voltage source (able to supply any desired current) in series with a resistance that limits the available current and causes the voltage across the battery terminals to drop.

The source resistance, also called the **internal resistance** or **source impedance**, is made up of the resistance of the metal connections inside the battery, and the effective resistance of the chemical reactions occurring inside the battery.

In a well-designed power supply the voltage will not drop as the current drawn is increased until the power supply reaches its maximum current. The effective source resistance is very low.

The voltage rating of a battery is the open circuit voltage; this is the potential difference across the terminals when there is no current flowing. This is also known as the **Electromotive Force (EMF)** of the battery.

When current is being drawn from the battery, the potential difference will drop as more current is drawn. This loss is due to the Internal Resistance or Source Resistance of the battery. The voltage drop can be calculated with Ohm's Law:

$$V_{drop} = I \times R_{source}$$

The voltage at the terminals can also be calculated:

$$V_{terminals} = V_{opencircuit} - V_{drop}$$

Capacitors

A capacitor is a component constructed from two metal plates separated by an insulator called the dielectric as is shown in **Figure 4.12** The capacitance is made larger by increasing the area, A, of the plates and decreasing the distance, d, between them. In this diagram the insulating material between the plates is air. There is a range of materials that can be chosen for the insulator depending on the circuit in which it is to be used. It's construction structure is hinted at in the circuit symbol, **Figure 4.13**.

Capacitance is the measure of how much charge a capacitor can store. The unit of capacitance is the **Farad**. One Farad is a large value; capacitors commonly encountered in radio circuits have capacitances ranging from micro-Farads (10^{-6}) to pico-Farads (10^{-12}). **Table 4.2** gives examples of how capacitance can be expressed in pico-, nano- and micro- Farads.

The construction of the capacitor, the area of the plates, and the thickness and **dielectric constant** of the insulator determine the capacitance; larger plates and a thinner insulator lead to higher capacitance. The dielectric constant of an insulator is a measure of how well it can turn an electric field into a stored charge.

Different dielectrics give different properties for the capacitor, air spaced capacitors tend to have a low capacitance and are often used in tuned circuits. Ceramic capacitors are physically small with higher capacitances in the tens to hundreds of pico-Farads.

An electrolytic capacitor consists of two sheets of aluminium foil wound in a coil. One of the sheets has a thin coating of aluminium oxide and between this and the other sheet of aluminium is an electrolyte gel. An electrolyte is a solution of chemicals which conducts electricity. They allow very high capacitance values in a physically small space, but with the disadvantage that the capacitor is electrically polarised – it must be connected to the circuit with its polarity in the correct direction.

There is a very important safety consideration when using electrolytic capacitors; they must be connected the correct way round otherwise there is a danger of their destruction, further damage to the equipment, and sometimes personal injury.

A potential difference connected across a capacitor will cause a current to flow into the capacitor as the plates charge up and store the supplied energy. After a time, no more current will flow and the capacitor is said to be charged. The charge persists even if the applied potential difference is removed. The stored charge can be dangerous, it will maintain the same potential difference across the terminals that the capacitor was charged with; this can persist for a long time. High voltage circuits should include a high value resistor across the capacitor terminals to dissipate the charge after a short while.

Capacitors in series & parallel

Just like resistors, capacitors can be connected in series or in parallel. The formulae look similar, but capacitors in parallel add their values together, and connecting capacitors in series reduces their capacitance.

Reading the value of a Capacitor

Regrettably there are many capacitor manufacturers and methods of writing the capacity and tolerance on the body of the component.

The simplest way of doing this and the one you need to know for your exam is that of using three numbers which give a value in pF. This is very similar to rating resistor values where the third number gives the number of 0s. Some examples are shown in **Table 4.3**.

When selecting capacitors you will need to ensure that you have the power rating and tolerance necessary for your circuit design.

Parallel capacitors

In **Figure 4.14**, you can see that the top two plates are connected together, and the bottom two plates are connected together. This is the same as replacing the two smaller capacitors, C_1 and C_2, with one capacitor of plate area the same area as the combined plate areas of C_1 and C_2.

The formula for the total capacitance of some number (n) capacitors in parallel is:

$$C_{total} = C_1 + C_2 + ...C_n$$

Example

A 100 µF capacitor is in parallel with a 33 µF capacitor. What is the total capacitance?

$$100\mu F = 100 \times 10^{-6} F$$

$$33\mu F = 33 \times 10^{-6} F$$

$$Total = 133\mu F (133 \times 10^{-6} F)$$

Series Capacitors

When capacitors are connected together in series, as in **Figure 4.15**, the total capacitance is reduced. You can think of the capacitors in series as being one capacitor with the plates spaced further apart. A smaller quantity of charge is stored on the capacitors – so the effective capacitance has been reduced.

The formula for *n* capacitors in series is

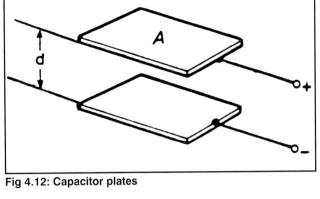

Fig 4.12: Capacitor plates

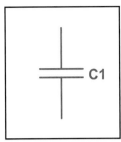

Fig 4.13: The schematic symbol for a capacitor represents its construction, two plates separated by a small distance

		F		mF		µF		nF		pF
1F	is	1	or	10^3	or	10^6	or	10^9	or	10^{12}
1mF	is	10^{-3}	or	1	or	10^3	or	10^6	or	10^9
1µF	is	10^{-6}	or	10^{-3}	or	1	or	10^3	or	10^6
1nF	is	10^{-9}	or	10^{-6}	or	10^{-3}	or	1	or	10^3
1pF	is	10^{-12}	or	10^{-9}	or	10^{-6}	or	10^{-3}	or	1

Example: We can see 1 Farad is:

1000 (10^3) milli-Farads, or

1,000,000 (10^6) micro-Farads or

1,000,000,000 (10^9) nano-Farads or

1,000,000,000,000 (10^{12}) pico-Farads

Table 4.2: Converting capacitor values between, F, mF, nF and pF

Number printed on the capacitor	First two numbers	Number of 0s	Value in pF
102	1 0	2	1,000
473	4 7	3	47,000
221	2 2	1	220
384	3 8	4	380,000
574	5 7	4	570,000
251	2 5	1	250
111	1 1	1	110

Table 4.3

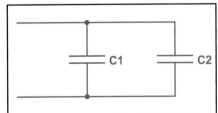

Fig 4.14: Two capacitors in parallel

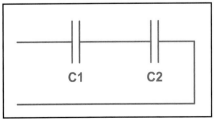

Fig 4.15: Two capacitors in series

$$\frac{1}{C_{total}} = \frac{1}{C_1} + \frac{1}{C_2} + \cdots \frac{1}{C_n}$$

Example
A 100nF capacitor is in series with a 22nF capacitor, what is the total capacitance?

$$100nF = 100 \times 10^{-9}F$$

$$22nF = 22 \times 10^{-9}F$$

$$\frac{1}{C_{total}} = \frac{1}{100 \times 10^{-9}} + \frac{1}{22 \times 10^{-9}}$$

So:

$$C_{total} = 18.03 \times 10^{(-9)} F = 18nF$$

Inductors

In 1831 Michael Faraday discovered that an electrical current in a wire produces a magnetic field. This is shown in **Figure 4.16.**

The wire is at right angles to and passes through a sheet of card. Using a small plotting compass and a pencil to mark its position we can show the lines of magnetic force around the wire. The point of each arrow on the card represents the North pole of the plotting magnet. This is summed up by the **Corkscrew Rule**: the direction of turning the corkscrew shows the direction of the magnetic line of force. Note the diagram refers to conventional current.

We need to look a little deeper into how we can make magnetic fields of practical use. A very common way is to wind the wire into a coil to increase the strength of the magnetic field. The magnetic fields round each turn of the coil add together to produce a much stronger magnetic field as in **Figure 4.17**

A simple coil will produce a much stronger magnetic field if the air is replaced by iron as shown in **Figure 4.18**

The energy from the electrical current is stored in the magnetic field as the current builds up when first switched on. This stored energy is released when the current ceases to flow.

When we switch off the current the magnetic field will collapse, it is changing again, albeit downwards. That will also induce a voltage in the wire, now trying (but failing) to keep the current going. The energy stored in the magnetic field is returned to the circuit.

This opposition to change is known as the Back EMF.

If there is a second wire in the same magnetic field voltages and currents will be induced in that wire too, every time the magnetic field changes. Some energy has transferred via the magnetic field from one wire to the other. It is helpful to keep this point in mind for when we look at transformers.

The idea of an interchangeability between current and magnetic field strength is used in making components called **inductors**. The ability of an inductor to store energy in its magnetic field is called **inductance** and the unit of inductance is the Henry, H.

Inductors are important components in Alternating Current Circuits and later we will pay particular attention to their property of **reactance** in **tuned circuits.**

Inductances used in radio circuits are often in the range of milli-Henry (mH) to nano-Henry (nH).

In general, the inductance of a coil depends on the number of turns, the core material used and the diameter of the coil.

To increase the inductance we would,
- Increase the number of turns.
- Increase the diameter of the coil.
- Use a core having a high permeability.

Permeability

Permeability is a word you do not need for the exam. Suffice it to say that air has a low permeability and iron and ferrite have a high permeability, in simple terms it is a measure of the ability to concentrate the magnetic field. You need to be aware that the core influences the value of the inductance and some substances have a stronger effect. There is an additional way in which we can vary the amount of inductance. By carefully adjusting the position of the core within the coil we can change the inductance. Thus, we have a variable inductance and this is shown **Picture 4.8.**

Radio circuitry is very much concerned with alternating currents. At low audio frequencies iron-core inductors may be used to concentrate the magnetic field. However, at higher frequencies, iron-dust or ferrite cores are used since iron cores become too lossy. In circuits using the higher frequencies such as VHF and UHF you may see air-cored inductors more frequently, because large inductance values are not required.

As with resistors and capacitors, you can connect inductors together in series and parallel circuits – although it is uncommon to connect inductors in parallel.

Series inductors

If you consider two inductors in series, it's quite easy to see that it is like one larger inductor with more turns, so the formula for *n* inductors in series is:

$$L_{total} = L_1 + L_2 + \ldots L_n$$

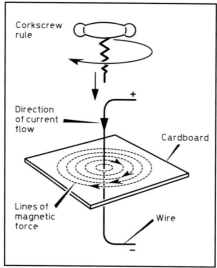

Fig 4.16: A magnetic field is produced by a current flowing in a straight wire

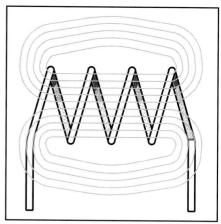

Fig 4.17: An Inductor and its magnetic field

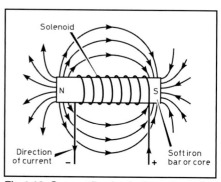

Fig 4.18: Current flowing in the coil creates a strong magnetic field

Parallel inductors
Inductors in parallel behave similarly to resistors in parallel. The formula for some number (n) inductors in parallel is:

$$\frac{1}{L_{total}} = \frac{1}{L_1} + \frac{1}{L_2} + \cdots \frac{1}{L_n}$$

Storing energy in Capacitors and Inductors
The formation of an electric field in a capacitor and magnetic field in an inductor does not happen instantaneously; it takes time. The following circuit and graphs will show us what happens.

Capacitors
In **Figure 4.19** when the switch is at position "b" there is no current flowing and the voltage across the capacitor is zero volts until the switch is moved to position "a". Current will flow to the capacitor and the voltage across the plates will rise to a steady value which is that of the battery. At this point there is no current flowing. The capacitor can be discharged by moving the switch to position "b". As is shown in **Figure 4.20** this will cause current to flow through the resistor and the voltage across the capacitor will decrease to zero.

Inductors
We can use the same circuit as above but replace the capacitor, C, by an inductor, L.

When the switch is at position "b" there is no current flowing (I= 0 Amps). Closing the switch to position "a" the current rises quickly and the magnetic field develops. As is shown in **Figure 4.21** the increasing magnetic field generates a back EMF which opposes the current and causes the rate of increase to reduce until it is zero. At this point the current has reached a constant value. When the switch is set to position "b" the magnetic field is no longer being maintained. As it collapses it induces a current in the coil which falls to zero.

Tolerance of Components
When designing circuits and choosing components we need to be aware that it is not always possible to get, for example, a resistor or capacitor, with the exact value that we want. Because of the manufacturing process there will always be small variations in their values. This variation is specified as its tolerance. A resistor stated to be 100Ω ± 1% will have a value between 99Ω and 101Ω. A capacitor rated at 100μF ± 20% will have a value between 80μF and 120μF. If the actual value is within this range we say that the value of the capacitor is within tolerance. Manufacturers will state the tolerance on the body of the component in writing or using a colour code. It helps to be aware that the value of a component may also vary with change in temperature and this also may be stated.

Alternating Current (AC)
You will know from your Foundation studies that electrical current may not always flow in one direction. In the case of alternating current, the current flow reverses every half-cycle. In the case of UK mains power, this change happens 50 times per second, or 50 Hertz.

Both alternating currents and direct currents will transfer energy. A filament light bulb will light up regardless of whether it is powered by AC or DC. Circuits that contain only pure resistances, such as heaters or electric kettles will behave the same on AC or DC, but their power rating will be different and we'll come to this later in this chapter.

In an alternating current, the rise and fall of the current and potential difference is normally shown graphically as a sine wave curve. There are other waves used in electronics such as square and triangle waves, but usually in radio circuits we are dealing with sine waves.

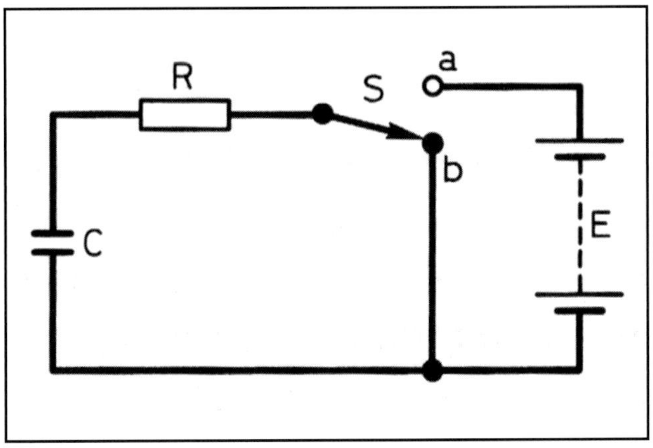

To describe the sine wave in a meaningful way we need to specify at least two parameters, the wave's amplitude and frequency. If dealing with more than one sine wave, it may also be useful to define phase.

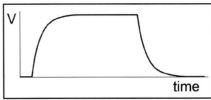
Fig 4.20: The change in voltage across a capacitor when it is charged and discharged

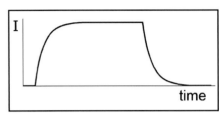
Fig 4.21: The change in current when an inductor makes and loses a magnetic field

Amplitude
The amplitude of the wave is the difference between the zero line and the peaks or troughs above and below it. The peak is more correctly referred to as a positive peak and the trough as a negative peak, which is usually a mirror image of the positive peak.

We can see from **Figure 4.22** that we could define another parameter of the sine wave, its peak-to-peak value. Not immediately obvious on Figure 4.22 is another measurement of the amplitude, the RMS value or Root Mean Square value. We shall see what these three ways of measuring amplitude actually mean.

The peak voltage is the difference between the top of the positive peak and zero, or the bottom of the negative peak and zero. As long as the wave is symmetric about zero these two values will be the same. This value is often labelled V_{pk}

Peak-to-Peak amplitude is the difference between the positive and negative peaks. A peak-to-peak value of potential difference would be labelled as V_{pk-pk}

AC Circuits, the values of Current and Voltage
In a DC circuit current and voltage have constant values and we can state their magnitude without difficulty. This is not so easy in an AC circuit where these values are constantly changing. We therefore have a challenge in giving a value to the amplitude of something which is in constant change.

Fortunately, there is a mathematical

Fig 4.19: Charging and discharging a capacitor

trick that allows us to do this. It is called the Root Mean Square value, RMS. You do not need to know anything about its derivation, only how to use it. It can be thought of as a way to turn the AC curve into straight line of equivalent DC value.

$$RMS = \frac{Peak\ Value}{\sqrt{2}}$$

Peak Value can be either Amps or volts and √2 means the *square root* of 2. This number is 1.414 and when multiplied by itself gives 2.

($\sqrt{2} \times \sqrt{2} = 2$ and 1.414 x 1.414 = 2)

The formula can be made easier for calculations,

$$RMS = \frac{Peak\ Value}{\sqrt{2}} = \frac{Peak\ Value}{1.414}$$

$$= Peak\ Value\ x\ 0.707$$

It is easy to remember that *the RMS value is 70.7% of the Peak Value.*

The RMS value becomes very important when we are looking at the power dissipated by AC and DC circuits. In a DC circuit there is a constant power output but in an AC circuit it changes between zero and maxima at the peaks. The use of RMS values gives us a bridge between ideas used in AC and DC circuitry.

The RMS current or voltage in an AC circuit is equal to the current or voltage of a DC supply that would result in the same power dissipation.

We can show this with a simple example. What are the current and voltage in a DC circuit having the same power output as an AC circuit in which the peak values are 28.28 Volts and 11.31 Amps?

The RMS voltage is
0.707 x 28.28 = 20 Volts
The RMS current is
0.707 x 11.31 = 8 Amps

The DC circuit will therefore be 20 Volts with a current of 8 Amps
Both the AC and DC circuit will be dissipating 8 x 20 = 160 Watts

UK Mains Voltage

It is important to note that the UK mains supply is given as an RMS value of 230 Volts. At the peak of the 50Hz cycle there is a maximum of 230 x 1.414 = 325 Volts.

Frequency

Along with amplitude (which ever measurement of it you use), the frequency of a single waveform is all you need to completely specify it. You will have encountered frequency before in the Foundation licence, so the following section is provided for completeness.

Frequency is measured in Hertz (Hz) and is the number of times an alternating current goes though its whole cycle (from zero, to a positive peak, back to zero, to a negative peak and again back to zero) in one second. Frequency is commonly given the symbol **f**. The UK mains electrical current alternates at 50Hz, while audio signals extend into the kilo-Hertz (thousands of times per second) range and radio waves extend into the giga-Hertz (thousands to thousands of millions of times per second) range.

As well as frequency, you may also see the period of a wave specified (given symbol **T**) which is measured in seconds. This is the time taken for one cycle of the wave. The formula that relates the period of a wave to its frequency is

$$\text{period (T)} = \frac{1}{\text{frequency }(f)}$$

On a graph showing a wave, you can measure the period by looking at how much time (on the horizontal axis) there is between one point on the sine wave and the next time that same point occurs – for example, the time between one positive peak and the next positive peak, or when the signal crosses the zero line going upwards and the next time it crosses the zero line going upwards.

You are expected to be able to convert between frequency and period, specifically to be able to determine the period of a wave from a graph, and from this periodic time, to be able to calculate the frequency of the wave.

Example
The UK mains frequency is 50Hz. What is the periodic time for this wave?
$$f = 50$$
$$\text{period} = \frac{1}{50} = 0.02\ \text{seconds} = 20\ \text{ms}$$

Example
A waveform is measured to have a period of 10 milli-seconds. What is its frequency?
$$T = 10\text{ms} = 0.01\ \text{seconds}$$
$$\text{frequency} = \frac{1}{0.01} = 100\text{Hz}$$

Wavelength

You may recall that the frequency of a radio wave is related to its wavelength. If you visualise the signal as a wave travelling through space, alternating a certain number of times per second (so many Hertz), the distance (measured in metres) from one peak to the next peak is one wavelength. This is shown in **Figure 4.23.**

For the Foundation exam you could use a chart or calculation for converting frequency in mega-Hertz to a wavelength in metres, and vice versa. You are now at a level where you can be asked to convert frequency to wavelength without a chart. Don't panic, it's not that difficult.

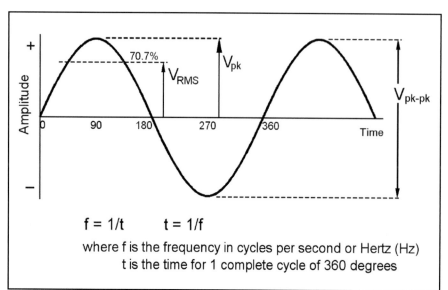

Fig 4.22: one and a half full cycles of a wave with the different measurements of amplitude marked

The link between frequency and wavelength is easy to understand if you remember that all radio waves travel at the same velocity (speed). If the wavelength of a radio wave is very long, you won't fit in so many cycles into one second. Conversely, if the wavelength is very short, you will get many cycles per second. This means that the longer the wavelength is the lower the frequency will be, and the shorter the wavelength is the higher the frequency will be.

Mathematically, we say that the velocity of the radio wave is equal to the frequency multiplied by the wavelength. Greek letter v (said like the word "new") is used to represent the velocity of radio waves. We have:

$$v = f \times \lambda$$

If you are good at maths you will know that you can rearrange equations so that if you know two of the three parts you can calculate the third, hence:

$$f = \frac{v}{\lambda} \quad \text{or} \quad \lambda = \frac{v}{f}$$

This is not a maths book, so let's make things a bit easier! If you can remember a simple triangle you can forget the equation and still get the answers. In the triangle shown in **Figure 4.24**, v is always on top. It is always the same number: 300. This 300 represents the velocity of the radio waves measured in millions of metres per second; it is 300 million metres per second.

Frequency (in MHz) and wavelength (in metres) are on the bottom. If you know the value of two of the numbers you can use these to calculate the third missing quantity.

Let's try a few examples:
- A transmitter is operating on a wavelength (λ) of 100m. What is its frequency?

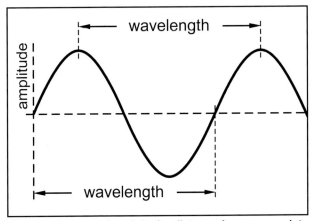

Fig 4.23: One wavelength is the distance from one peak to the next

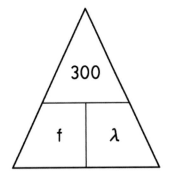

Fig 4.24: Magic triangle relating frequency and wavelength

You don't know the frequency, so you cover 'f' on the triangle. What remains is 300 over λ or 300 divided by λ. In this example, 300/100 = 3MHz. Using 300 for v always gives the answer in MHz.

- A receiver is tuned to frequency of 1.5MHz. What is the wavelength of the received signal?

This time we cover λ which leaves us with 300/f. In this case 300/1.5 = 200m. Remember, if you use 300 on the top and f is in MHz, the wavelength will be in metres.

- Here's a trickier question: What is the wavelength of a signal at 10GHz?

We must cover λ, as we don't know the wavelength. That leaves 300/f, but in this example, f is not in MHz, so you need to convert it. 10GHz is equal to 10,000MHz (remember the extra three zeros). The triangle now reads 300/10,000 = 0.03m or 3cm, a very short wavelength.

Try calculating the wavelength of the 0.136MHz band, and then decide on the lengths of wire needed to build a half-wave dipole! If you get stuck, ask your instructor or a more experienced amateur.

Radio amateurs often refer to the various bands allocated to them in terms of metres, because wavelength is an important concept. For example, you might hear amateurs talk of the 40m band, which refers to the 7MHz band. If you do the calculation, you discover that 40m is more correctly 7.5MHz (not 7MHz), so the name 40m band is not technically correct but it is nonetheless used throughout the world. The same could be said of 145MHz. Is it really 2m?

Phase

If we have two or more waves they can differ in amplitude, frequency and phase. If the two waves have the same frequency and start at the same time, then they are in phase. If one of the waves is delayed with respect to the other, they are no longer in phase.

Phase is measured in degrees and is based on there being 360 degrees in a full wave cycle.

In **Figure 4.25**, the two waves A & B are in phase but have different amplitudes. Wave C is not in phase with A, it leads A by a quarter of a cycle or 90°. Wave C leads because it reaches its maximum amplitude before waves A and B.

Harmonics

If two waves have different frequencies, but the frequencies are exact multiples of each other, for example the second frequency is at 3 times the frequency of the first, then the second wave is a **harmonic** of the first. It is important to note that harmonics can be generated inside oscillators and power amplifiers and may be at frequencies outside the amateur bands, so they must be filtered out.

Example

If you are transmitting on 51.510MHz (the 6-metre band FM calling frequency), then your second harmonic will be at twice this frequency, at 103.02MHz. This frequency is in the FM broadcast band.

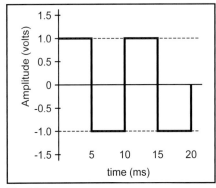

Fig 4.26: a square wave

Square Waves

As well as recognising a sine wave you need to know what a square wave looks like. The amplitude of the wave alternates at a steady frequency between fixed maximum and minimum values. In the Full Licence course you will learn of their use in digital circuits.

4: Basic Electronics

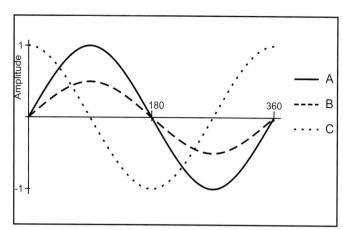

Fig 4.25: Three waves, with differing amplitudes and phase

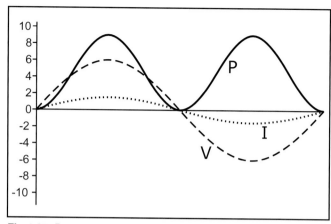

Fig 4.27: Power, current and voltage in an AC circuit

Capacitors, Inductors, Resistors and Alternating Current

So far, we have considered capacitors, inductors and resistors either alone or as part of small circuits and we have only looked at their properties when a direct current has been applied to them. This is not how they are commonly used; both audio and radio circuits will deal with alternating current – currents that vary their amplitude with time and have particular frequencies.

Resistors

A perfect resistor will behave no differently if it is passing a DC or an AC current, it will present its resistance and nothing more.

Ohm's Law still applies for an alternating current in a resistor. The current and voltage are in phase; they rise and fall together. This is shown in **Figure 4.27** which you need to be able to recall.

Of course, there is no such thing as a perfect resistor, the leads to the resistor act like low value inductors for example. For the purposes of the Intermediate licence we can consider resistors to only offer their marked resistance independent of their usage in AC or DC circuits.

Power in a resistor

Figure 4.27 shows the regular rise and fall of current and voltage in an AC circuit with a resistor. We recall that Power = Amps x Volts. Consequently we can see that where current and voltage have negative values, power will show a positive peak.

$$(-I \times -V = +P)$$

Capacitors

When connected to a voltage source, a charge current will flow into the capacitor until the potential difference across the capacitor is the same as the voltage source it is connected to; we say the capacitor is charging. Once the capacitor is charged, no further current can flow in the circuit. From the point of view of a DC current a charged capacitor appears to be an open circuit; no current can flow between the terminals once the capacitor is in this state.

The only way to get current to flow again is to discharge the capacitor. This can be achieved by connecting both leads of the capacitor together. Once discharged the current stops and can only flow again if the capacitor is once again charged.

In an AC circuit the polarity of the current is regularly alternating, and so provides a continual charge-discharge cycle, that allows a current to flow though the capacitor. The voltage across the capacitor is out of phase with the current though the capacitor, i.e., the voltage and current do not reach maximum at the same time. In a capacitor, the current leads the voltage by 90 degrees, as shown in **Figure 4.28**.

The flow of alternating current though a capacitor is governed by the value of the capacitor (measured in Farads) and the frequency of the applied alternating current. High frequency currents will pass through a low value capacitor relatively easily and we say the *reactance* to current flow is low (*reactance* is explained opposite). At lower frequencies the reactance is high and conduction of current is much reduced. That is because at a lower frequency the time for each half cycle is longer and the capacitor has to store more charge, which is harder to do.

An important point to remember is that by repeatedly charging and discharging in alternate directions a capacitor can pass alternating currents but not a direct current.

Inductors

Inductors have the opposite behaviour to capacitors when presented with an alternating current: A low frequency current will pass relatively unimpeded through a low value inductor, but a high frequency current will be impeded.

As shown in **Picture 4.4** an inductor's construction is just a length of wire formed into a coil, and so there is a continuous path for direct current to flow, perhaps with some resistance due to the natural resistance of the wire.

You may note that the amount of inductance can be increased by,

- a greater number of turns
- an increased diameter, d
- a core made of iron

Current flow through an inductor causes the formation of a magnetic field around the inductor. If the current continues, the magnetic field will increase to a particular strength depending on the current, the number of turns of wire, and the material of the inductor core, then stabilise.

When the current flow stops the mag-

Picture 4.4: An Inductance Coil with an air core.

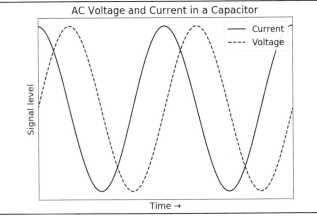

Fig 4.28: Current lead in capacitor

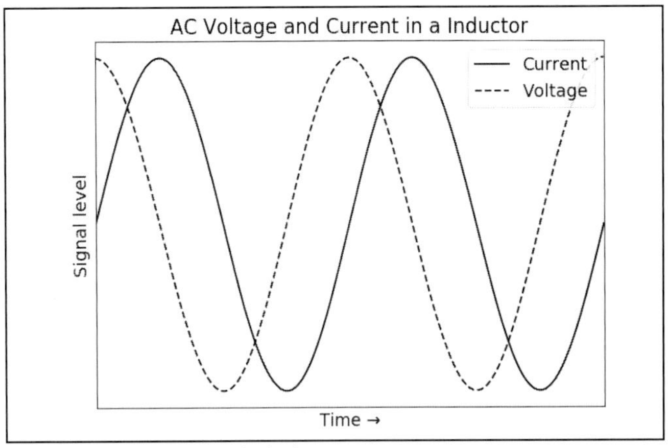

Fig 4.29: Voltage lead in inductor

netic field begins to collapse and causes a current to flow in the coil. This was discussed in **Storing energy in Capacitors and Inductors** earlier in this chapter.

If an alternating current is applied to the inductor, by its very nature the applied current direction changes every half-cycle. This causes the magnetic field to grow and collapse continually. The collapse of the magnetic field happens quite slowly and hinders the growth of the next cycle of the magnetic field. The details of this behaviour are discussed further in the Full licence book, but for now it is enough to understand that this slowness in field collapse results in the inductor allowing lower frequency (slower changing) currents to pass through while blocking higher frequency (faster changing) currents.

In an inductor, the voltage across the inductor leads the current though it by 90 degrees, as shown in **Figure 4.29**.

At VHF and above, even short lengths of wire can have significant inductance and stop circuits behaving exactly as expected.

The opposition to AC current flow through capacitors and inductors is called **Reactance**, and is measured in Ohms, the same unit as resistance.

Reactance and Impedance

Reactance is the AC equivalent to resistance in DC circuits. For a given component it's reactance will depend on the frequency of the current applied to it. The symbol given to reactance is X and the units it is measured in are Ohms.

Capacitive reactance

Having a layer of insulator between its plates means a capacitor cannot pass a DC current; it can pass an AC current because the plates rapidly charge and discharge as the direction of the AC current changes, as we have seen.

A capacitor will not pass all frequencies equally. A capacitor will more easily pass higher frequencies than lower frequencies. The reactance of a capacitor is given the symbol X_C.

Capacitive reactance falls as the frequency of the AC applied increases, this can be seen in **Figure 4.30**.

Inductive reactance

An alternating current flowing through an inductor creates a varying magnetic field. This changing field tends to oppose the current direction as we have seen previously, and resists changes to the current flow. This is inductive reactance and is given the symbol X_L

The inductive reactance increases as the frequency of the AC applied increases, this is shown in **Figure 4.31**.

It is important that you are able to recall, that reactance is the opposition to current in a purely inductive or capacitive circuit where the phase difference between V and I is 90°

Determining the value of Reactance

We have said that reactance is the AC equivalent of resistance in a DC circuit. Recalling the formula for Ohm's Law,

$$R = \frac{V}{I}$$

and looking at the phase relationship between I and V in **Figure 4.32** we have an unexpected problem.

At time "A" the current is measurable but V = 0 suggesting R is zero

At time "B" the voltage is measurable but I = 0 suggesting R is infinite

The way in which we overcome this is to use the RMS values of V and I

$$X_C = \frac{V_{RMS}}{I_{RMS}}$$

Similarly, we can see that

$$X_L = \frac{V_{RMS}}{I_{RMS}}$$

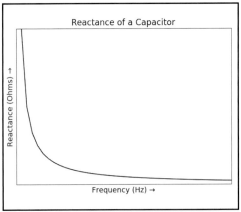

Fig 4.30: The reactance of a capacitor vs frequency

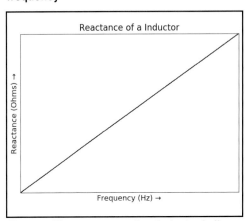

Fig 4.31: The reactance of an inductor is linear with frequency

4: Basic Electronics

Impedance

The combination of **resistance** and **reactance** in a circuit is called **impedance**. It, too, is measured in Ohms and is given the symbol Z. The impedance of a circuit comes from the way capacitors and inductors store energy in their electric or magnetic fields and the way resistors convert some of the energy into heat. Impedance is the opposition to both energy storage and transfer in a circuit.

While resistance and reactance are both measured in Ohms, you cannot simply add the values together, since we must account for the phase differences between voltage and current through reactive components. You will learn more about the special way reactance and resistance are combined in the Full Licence course.

You are, however, required to know at Intermediate level that the ratio of the overall supply voltage to current is the impedance, Z, of the circuit,

$$Z = \frac{overall\ supply\ voltage}{current}$$

Transformers

You will probably already be aware of transformers and their use in power supplies. You may know that power transformers are formed from two or more coils of wire around an iron core.

The primary coil is connected to a source of alternating current which generates an alternating magnetic field, as we have seen when discussing inductors. The magnetic field is conducted through the core of the transformer which passes through the centre of the secondary coil. The changing magnetic field in the core of the secondary coil generates an alternating current in the secondary coil winding. as shown in **Figure 4.33** Depending on the transformer design, there may be more than one secondary coil, but all the secondary coils share the magnetic field created by the primary coil. **Picture 4.5** shows a small transformer built into a domestic power supply.

The purpose of the iron core is to concentrate the magnetic field. Iron cores are not suited to frequencies greater than a few hundred Hertz as they heat up and waste power. For radio frequencies, cores composed of ferrite (a ceramic material containing iron) or compressed iron dust are commonly used; these materials are capable of working over wide ranges in frequency with much better efficiency than an iron core.

If the secondary coil has fewer turns than the primary coil, the transformer is a step-down type, and the voltage on the secondary coil will be lower than the voltage on the primary coil. In the opposite case, where there are more turns on the secondary coil than the primary, the transformer is a step-up type, and the voltage on the secondary coil will be greater than the voltage on the primary.

It is important to note that transformers only work with alternating current. If a direct current is applied to the primary coil of a transformer, it will generate a constant (non-changing) magnetic field. Since this magnetic field does not change, it will be incapable of inducing a current in the secondary coil. The output of a transformer fed with AC will be AC.

Tuned circuits

An inductor paired with a capacitor forms a **tuned circuit**. There are two ways of connecting the components as shown in **Figure 4.34**:

- In series to form a series tuned circuit
- In parallel to form a parallel tuned circuit.

To see how a tuned circuit operates, we need to understand what happens to the current in the circuit. Consider the parallel tuned circuit for this explanation as it is somewhat easier to visualise.

If we connect a charged capacitor in parallel with an inductor, current will flow from the capacitor into the inductor producing a magnetic field though the inductor's coil.

The capacitor discharges and the potential difference across the circuit drops to zero; this causes the magnetic field in the inductor to collapse and feed energy back into the coil inducing a current. This current recharges the capacitor until all the energy stored in the inductor's magnetic field has been returned to the capacitor. The transfer of energy between the capacitor and the inductor is continuous and carries on until all the energy has been dissipated due to losses in the circuit.

The result is an oscillating current between the inductor and capacitor. The number of times per second the current changes direction is the circuit's resonant frequency. At the resonant frequency, the capacitor and inductor act together to control the flow of current in the circuit.

The energy stored in the capacitor and inductor in a tuned circuit is transferred from one to the other at the resonant frequency.

If an alternating current is applied to the tuned circuit with a frequency away from the circuit's resonant frequency, the oscillation of energy between the capacitor and inductor does not occur. This is somewhat analogous

Picture 4.5: The contents of a small domestic power supply. The transformer is clearly visible on the right hand side. The primary and secondary coils are wound within interleaved iron strips.

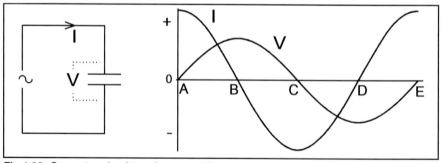

Fig 4.32: Current and voltage in a capacitor.

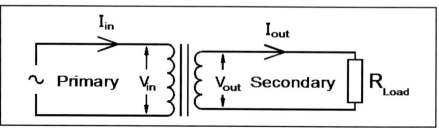

Fig 4.33: Circuit representation of a transformer.

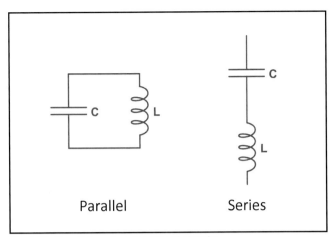

Fig 4.34: Parallel and series tuned circuits

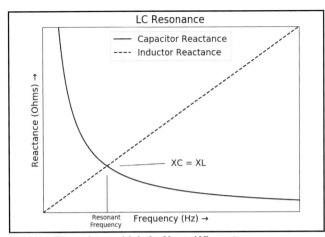

Fig 4.35: The point at which the Xc and Xl reactance curves cross is the resonant frequency of the circuit

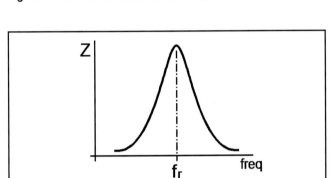

Fig 4.36: Parallel resonance

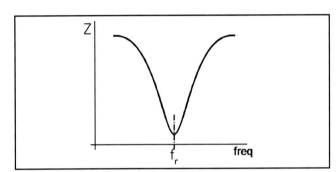

Fig 4.37: Series resonance

to pushing a child's swing at the wrong time.

The resonant frequency is determined by values of the capacitor and inductor, changing either of these will change the resonant frequency. Reducing the value of the capacitor or inductor leads to an increase in the output frequency. Similarly, increasing the value of the inductor or capacitor lowers the output frequency. It is possible to lower the capacitor value and increase the inductor value (or vice versa) and maintain the same resonant frequency.

A tuned circuit with a resonant frequency at 7.1MHz can be tuned to 7.2MHz either by reducing the capacitor value or reducing the inductor value. Often such a circuit in a radio will have a fixed inductor and a variable capacitor to allow the frequency to be tuned.

At resonance, the capacitive and inductive reactance values are equal. You can see this as the point where the two plots, X_L and X_C cross. This point is at the resonant frequency as shown in **Figure 4.35**.

Current flow in series and parallel tuned circuits

The two combinations of capacitor and inductor, series or parallel, as shown in the diagram in **Figure 4.34** behave differently when you consider how an alternating current flows through them.

The components used in real circuits are never perfect, so this must be factored into consideration; both the inductor and capacitor have some loss, usually wire resistance in inductors and dielectric losses in capacitors.

The parallel tuned circuit may be the one you're most familiar with, especially if you've built crystal sets or other simple radios. Parallel tuned circuits are often used in the main tuning of these types of radios.

The parallel tuned circuit presents a high impedance at its resonant frequency and a low impedance above and below that frequency. In simple radios this property of the circuit is used to select which frequency will be passed along to the rest of the receiver, all other frequencies will pass though the tuned circuit to ground and cause no signals in the rest of the radio.

Another common use of the parallel tuned circuit is in *traps* for beam or dipole antennas. The tuned circuit effectively isolates the extra length of the antenna elements at the resonant frequency. Several traps can be used to form multi-band antennas. This is discussed more in the Antenna Concepts chapter.

The series tuned circuit presents a low impedance at its resonant frequency. These circuits often find use in filters between stages in radio circuits. The frequency that should pass between stages does so with low impedance, while all other frequencies see a high impedance and are attenuated.

You need to know and recognise how the impedance of parallel and series tuned circuits varies with frequency. The series tuned circuit shown in **Figure 4.37** presents a low impedance and the parallel circuit shown in **Figure 4.36** presents a high impedance. Satisfy yourself that graphs of current plotted against frequency will have the inverse shape of curve

Figure 4.38 shows some simple circuit configurations for low-pass, high-pass, band-pass and band-stop (notch) filters. You will need to identify these circuits and their response curves.

Specifying the value of high and low pass filters

You will see later when looking at Q factors we are concerned with the frequency at the half power point. High and low pass filters are no different. The frequency at the half power point is known as the cut off frequency. You will notice the vertical axes in **Figures 4.39** and **4.40** of the two graphs are labelled voltage and not as we might expect power. The reason for this is that voltage is easy to measure. Earlier you learnt that power is proportional to V^2. With a little manipulation of

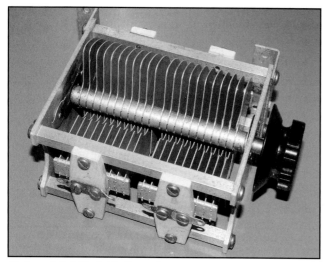

Picture 4.6: A Variable Capacitor

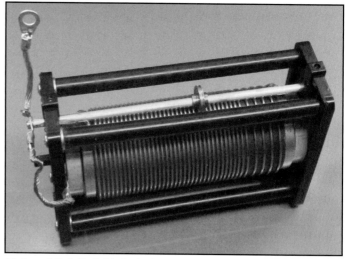

Picture 4.7: A Variable Inductor

Picture 4.8: A Slug Tuner

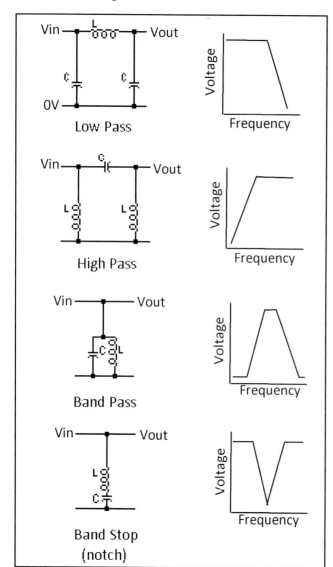

Fig 4.38: Common filter topologies and their frequency responses

the formula we can show that if a voltage of 1 Volt is reduced to 0.707V the power has been halved; it has been changed by -3dB (The dB or decibel is explained in Chapter 11)

The high pass filter has a low voltage output at the lower frequencies. It significantly attenuates them up to the -3dB power point and allows the higher frequencies to pass.

The low pass filter has a high voltage output at the lower frequencies and at the -3dB power point the higher frequencies are greatly attenuated.

Components needed for tuning

You will now realise that in a tuned circuit it is essential that we are able to vary the amount of inductance or capacitance. Fortunately we have variable capacitors and variable inductors.

A **variable capacitor** as shown in **Picture 4.6** has a centre spindle that when rotated moves the vanes such that the area of overlap between them decreases and the capacitance decreases.

Picture 4.7 shows a **variable Inductor** which is sometimes called a roller coaster, the inductance varies as the centre spindle is turned and the rider moves its position of contact along the coil.

With the slug tuner in **Picture 4.8**, a ferrite core is moved in and out of a coil as it is turned on a thread. This causes the inductance to change and is used in slug tuning.

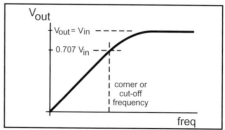

Fig 4.39: A high pass filter output

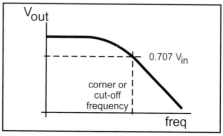

Fig 4.40: A low pass filter output

5: Semiconductors

Materials that conduct electricity do so because they are composed of atoms which have loosely attached electrons. Applying a DC voltage across a conductor causes the electrons in the conductor to drift towards the positive end. It is this drift of electrons that we call current.

Electrons have a negative charge.

As well as conductors (such as copper) that pass electrical current easily, and insulators (like glass and plastic) that prevent the passage of electric current, there exists a class of materials that have properties somewhere between conductor and insulator – these are referred to as semiconductors.

The most commonly used semiconductor material is silicon. On its own, pure silicon isn't very useful; it has a fairly high resistance. To make useful semiconductors, the pure silicon has to be 'doped' with other elements that either donate spare electrons to the silicon, or have fewer electrons leaving the silicon with a shortage of electrons called holes. Silicon doped with electron donating elements is referred to as an N-type material, and silicon doped with elements that form holes is referred to as a P-type material.

P-type semiconductors have almost no free electrons to conduct current, instead they contain holes. A hole is the absence of an electron. N-type semiconductors have an excess of electrons that can easily move though their structure and carry current.

When N-type and P-type semiconductors are joined together, they form a semiconductor junction or P-N junction; this is the basis of all the diodes, transistors, LEDs, and other semiconductors we use.

Diodes

One of the simplest semiconductor devices, the diode, consists of a piece of N-type semiconductor joined to a piece of P-type semiconductor to form a P-N Junction.

When a P-N junction is formed, some of the electrons in the N-type region diffuse across the junction to the P-type region where they "fill-in" some holes. This leaves a thin layer at the junction that is deficient in both holes and electrons, and

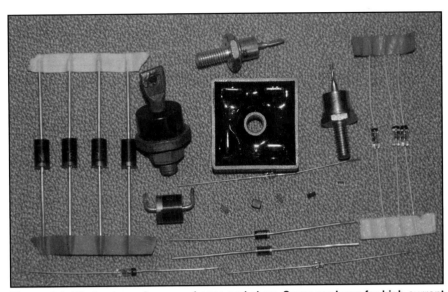

Picture 5.1: Diodes come in many shapes and sizes. Some are large for high current while others are small for weak signal rectification

so carries neither positive nor negative charge; it's an insulator. This region is called the **depletion layer**, and is responsible for many of the useful properties of the P-N junction.

To understand how a diode works it is helpful to have a clear understanding of what we mean by the term current; it is the movement of negatively charged electrons being attracted to something with a positive charge. In a circuit the electrons flow from the negative to the positive terminal of a cell. This makes sense as unlike charges attract each other. Regrettably, this was not fully understood by early physicists who wrongly asserted that current flows from positive to negative. This belief has been maintained by the use of the term conventional current.

Electrons can only flow through the diode in one direction, from the N-type side to the P-type side. When a potential difference is connected with the correct polarity across the junction the N-type semiconductor injects electrons through the depletion layer into the P-type semiconductor enabling it to conduct. When the polarity is reversed no current can flow because the P-type semiconductor has no spare electrons it can send into the N-type semiconductor. We have a device that only conducts current in one direction, the **diode**. The nature of the conventional current flow is shown by the arrow in the diode's circuit symbol, **Figure 5.1**. Diodes come in many shapes and sizes, depending on their intended use, as can be seen in **Picture 5.1**.

Even with the correct polarity across the diode, a current will only flow when the potential difference reaches a certain value that depends on the exact nature of the junction and the semiconductor materials used to make it. For common silicon diodes, the potential difference needs to be around 0.6 Volts before any current will flow – this is known as the **forward voltage** of the diode. This corresponds to the energy needed to inject electrons through the depletion layer.

Another type of diode sometimes used in radio receivers is the **germanium** diode; this uses another semiconductor material,

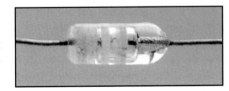

Picture 5.2: A Diode. It is encased in glass and is about half an inch long

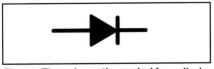

Fig 5.1: The schematic symbol for a diode; the arrow shows the direction of conventional current.

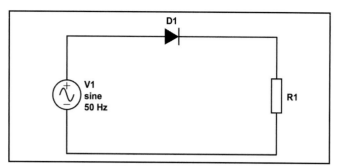

Fig 5.2: Circuit diagram for a single diode half-wave rectifier

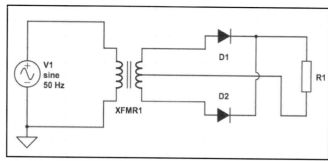

Fig 5.4: A full wave rectifier using a centre-tapped transformer and two diodes

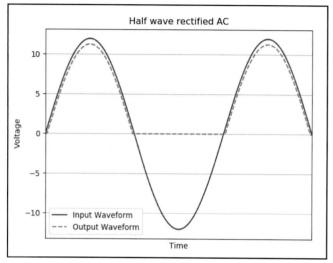

Fig 5.3: The input and output waveforms of a half-wave rectifier

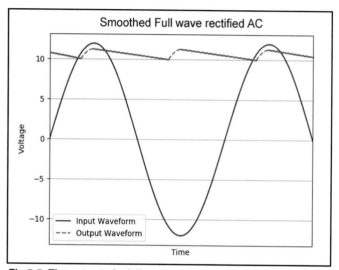

Fig 5.5: The output of a full wave rectifier with a large smoothing capacitor on the output

germanium, to produce a diode with a much lower forward voltage (around 0.25 Volts), these diodes are used in the circuits that detect the radio signals, and so have to work with very low voltages.

Rectification

Diodes are used in power supplies because of their unidirectional properties. They are used in the process of turning AC into DC, called rectification. You will cover this in more detail in the Full Licence course, but for now we consider two types of rectifier and their advantages and disadvantages. You will note that in **Figures 5.2, 5.4** and **5.6** the circuit load is represented by R1.

Half-wave rectification

A single diode connected between an AC supply and a load will cause an output as shown in the graph in **Figure 5.3**. This is half wave rectification. You can see that the negative going half of the AC voltage is clipped off, and the result is a series of pulses, their amplitude is lower than the input AC due to the forward voltage of the diode.

Half wave rectification is simple, but not very efficient. At least half the energy in the incoming AC is unused – it never makes it to the output. It is very rare to see a half wave rectifier in use in power supplies because of this inefficiency.

Full-wave rectification

By adding another diode to the circuit, we cause the negative portion of the AC current to be flipped to the positive side, immediately doubling the number of pulses on the output, and hence improving the efficiency.

The transformer in Figure 5.4 provides step down from the AC mains and is also centre-tapped on the secondary.

While these circuits are simple, the output from them isn't very useful. Instead of an alternating current swinging between positive and negative limits, we have a pulsed current that periodically becomes zero. Some way to "fill-in" the space between the pulses would help make some-

thing much closer to a DC supply. We can make this happen if we connect a capacitor across the output of the rectifier, in parallel with the load, as shown in Figure 5.6. The value of the capacitor in a typical circuit will be hundreds to thousands of micro-Farads. This capacitor is commonly known as a smoothing capacitor or reservoir capacitor.

You can see from the plot of the output of the smoothed full wave rectifier (**Figure 5.5**) that the output looks a lot more like a constant DC supply. The voltage still fluctuates a little, but never becomes zero. Comparing this with the unsmoothed output (**Figure 5.7**) you can see what is occurring; the capacitor charges when the voltage pulses rise from zero, and dis-

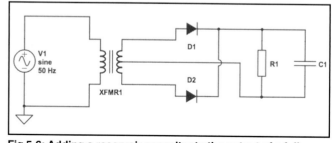

Fig 5.6: Adding a reservoir capacitor to the output of a full wave rectifier

Intermediate Licence Manual

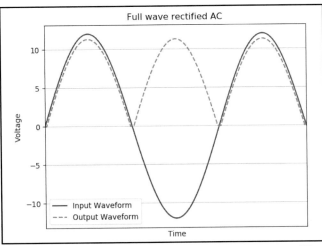

Fig 5.7: The output of an un-smoothed full-wave rectifier. The output is a series of pulses at twice the frequency of the incoming AC

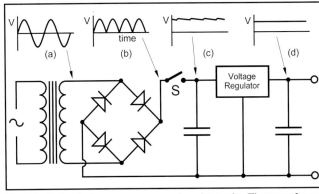

Fig 5.9: A simple linear power supply schematic. The waveforms show the voltage changes up to the indicated points.

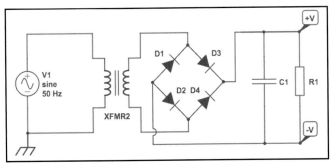

Fig 5.8: A bridge rectifier. The positive supply comes from the junction of D3 and D4, the negative comes from the junction of D1 and D2

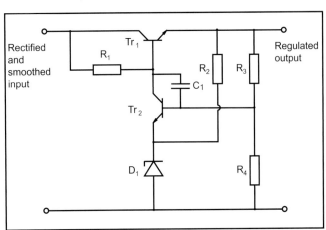

Fig 5.10 A voltage regulator which is often included as a separate integrated circuit.

charges into the load as the voltage of the pulses falls. This keeps the voltage at the load fairly constant. If the capacitor value was increased further, then the output would be even smoother.

Bridge Rectifier

The full wave rectifier with two diodes works well but relies on a centre-tapped transformer for its operation; this is not always convenient or possible. A way to overcome this is to use a bridge rectifier, this uses four diodes (D1-D4) arranged as shown in the circuit in **Figure 5.8**.

The operation of the bridge rectifier is complex and not required for the Intermediate licence, but is covered in the Full licence. For the Intermediate licence you will only be expected to identify the type of rectifier from its circuit diagram and be able to identify AC, pulsed and smoothed DC on graphs of rectifier output.

The output of the bridge rectifier is the same as in the full wave rectifier, and the same smoothing capacitor arrangement is used on the output.

Bridge rectifiers are the most common type used in power supplies, either constructed from discrete diodes wired together, or from commonly available pre-package modules as seen in the centre of Picture 5.1.

There are occasions when it is necessary to have a power supply that maintains a constant voltage. By looking at **Figures 5.9** and **5.10** we are able to see how this can be achieved.

In the first instance we will look at Figure 5.9 when the switch is open. The transformer reduces the input voltage and the inset graph, (a), gives the expected alternating positive and negative voltage peaks.

The next stage uses a bridge rectifier to give a series of positive peaks as shown in graph (b). Up to this point the right hand side of the circuit has not been connected and no voltage would have been shown in graphs (c) and (d). Upon closing the switch the capacitor smooths the series of humps to an almost straight line as is shown in graph (c) with only minor changes in voltage indicated by a slight slight ripple.

You will note that the thin arrows from graphs (b) and (c) are pointing to the same connecting wire in the circuit. In effect, they are connected to the same point in the circuit This means that with the switch closed graphs (b) and (c) will be identical.

The final step in the process is the action of the voltage regulator to produce the straight line graph, (d).

Figure 5.10 shows the circuitry inside the Voltage Regulator. It is designed to ensure that whenever the input that whenever the input voltage varies the output voltage remains constant. It uses two transistors and a Zener Diode as in Figure 5.10.

Tr_1 is referred to as a **series pass transistor**. It is in series with the input and output voltages and handles the entire load current. Tr_2 is used to measure the output voltage from Tr_1 and in conjunction with the Zener diode (which you will meet in the Full Licence material) ensures the output voltage remains constant.

Voltage regulators are commonly needed for a variety of electronics applications and they are often included as a separate integrated circuit.

You are not expected to know the circuits in Figs 5.9 and 5.10 but you should be able to recognise them, remember their purpose and identify their components.

Linear regulated supplies are not very efficient, produce a lot of heat and require heat sinks which make them heavy and

5: Semiconductors

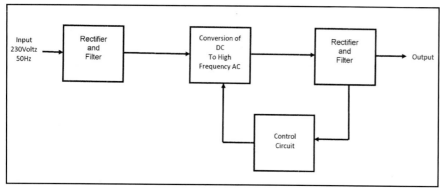

Fig 5.11: A Switched Mode Power Supply (SMPS)

expensive. Thirdly, there is a switched mode power supply. This is dealt with in greater detail in the Full Licence but you do need to recognise its circuitry and it is helpful to have an appreciation of how it works. The sequence of steps is,

- Incoming AC voltage is rectified and
- converted into a high frequency AC square wave, stepping the voltage up or down followed by
- rectification to DC and filtering and
- voltage regulation is achieved in the control circuit

The advantages of switched mode power supplies are high efficiency, little heat is produced, they are compact and relatively inexpensive. They do, however, produce a lot of high frequency harmonics and very effective screening and filtering are required to prevent RF interference.

Transistors

If a junction between two semiconductors produces the diode, you might ask how two junctions would behave. We can produce such a device by sandwiching a piece of P-type semiconductor between two pieces of N-type semiconductor, or with a piece of N-type between two pieces of P-type. Such an arrangement is known as a **bipolar junction transistor (BJT)**.

Picture 5.3 shows a selection of transistors commonly available.

Bipolar junction transistor (BJT)

The two types of transistors are known by the arrangement of the materials inside them, NPN or PNP (**Figure 5.12**). The Intermediate exam only considers the more common NPN transistor.

Wires attached to the semiconductor layers allow us to wire the transistor into a circuit. The wire attached to the middle layer, P-type in an NPN transistor, is called the **base (B)**. The wires attached to the other two N-type layers are called the **collector (C)** and the **emitter (E)**.

In an NPN transistor, a current can only pass from the collector to the emitter when a current is also passing from the base to the emitter. The base current is typically much smaller than the collector current. Since the base-emitter junction looks like a P-N junction of a diode as seen previously, it also incurs the same forward voltage drop of around 0.6 Volts when forward biased.

This small base current can allow a much larger collector current to be controlled. This is the basis of the **transistor switch** and many other types of amplification.

Other Transistor Types

Aside from the bipolar junction transistor discussed previously, there are many other types of transistor. One of particular note is the **field effect transistor (FET)**. These use the same N-type and P-type materials, with a slightly different arrangement. The terminals on a FET are called the **source, gate** and **drain**. FETs are not examined in the Intermediate licence, but you will come across them in the Full Licence Manual.

Integrated Circuits

It is possible to create many interconnected transistors, diodes and resistors on the same piece of a semiconductor material. This is most commonly silicon and technically known as the substrate. Such collections of components are called Integrated Circuits or ICs for short. These can be designed to perform complete circuit functions such as those found in amplifiers, oscillators, voltage regulators and digital processing systems.

Some common ICs you may encoun-

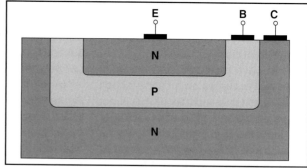

Fig 5.12: Diagram of internal construction of transistor

Picture 5.3: A selection of commonly available transistors

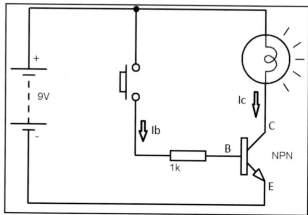

Fig 5.13: A simple transistor switch schematic

ter include the LM386 audio amplifier, the NE602 oscillator/mixer, as well as microprocessors such as the Atmega328p used on Arduino development boards common with hobbyists.

Transistors in circuits

The transistor switch

Transistors are commonly used in digital circuits as an electronic switch, in either an 'on' or 'off' state. It is used when you need to control a larger output current with a small control signal current.

Figure 5.13 shows an example of a transistor being used as a switch. When the push-button is pressed, a base current, Ib, flows through the 1kΩ resistor into the transistor base and out through the emitter back to the battery. In turn, this small base current controls a larger collector current, I_c, powering the lamp. The base current in this example is around 8mA, but the collector current could be several Amps! For this to work reliably, the transistor must be turned fully on (saturated), so the voltage between the collector and emitter is close to zero and the collector current is limited only by the load – in this case, the lamp.

The transistor amplifier

The transistor switch works well when the input signal represents an 'on' or 'off' state. To linearly amplify (produce a larger copy of the input signal at the output, without adding distortion) the transistor must be biased in the linear region. **Figure 5.14** shows a simple circuit that achieves this.

Resistor R1 sets the base current, into the transistor, while resistor R2 acts as the load. The voltage should be set at half the supply voltage to allow the maximum linear voltage swing – it is important that the output voltage does not go to the supply voltage or to ground, as distortion will occur (clipping). To function correctly, the transistor must be presented with the correct DC voltages and currents – this is known as **biasing**.

Figure 5.14 shows the circuit for how a transistor acts as an amplifier. The Capacitors C1 and C2 are important in that they block external circuits which may affect the biasing and are selected on the amplifier's frequency of use. They are not relevant to showing how amplification and gain are explained.

When the incoming signal, or input, I_b is positive (i.e. a peak) more base current will flow and when the signal is negative (i.e. a trough) less base current, I_b will flow.

This incoming AC signal, to the base of the transistor, I_b is exactly copied by a larger current flowing between the emitter and collector I_c. This **amplification** or **gain** is given the symbol β.

If I_c is 100 times bigger than I_b the gain, β, is 100. We can say this mathematically as

$$\beta = \frac{I_c}{I_b}$$

You may see β expressed another way by the letters h_{fe}. You need not concern yourself with this but you should be able to use the equation to determine β, I_b and I_c when given the values of two of the other symbols.

For example, if the value of R1 has been chosen to give a current, I_b, of 10μA and β is 100 we can calculate the collector current,

$$\beta = \frac{I_c}{I_b}$$

which can be rearranged to give
$I_c = \beta \times I_b$
which is $10^2 \times 10 \times 10^{-6} = 10 \times 10^{-4}$ A
which is 1×10^{-3} A or 1mA.

Although it is not required for your exam you may be interested in circuit design and will see that we are now in a position to calculate the value of R2. We have said that the voltage at the transistor should be half the supply voltage. Hence, 12 / 2 = 6 volts across the transistor and 6V across R2. Using the above example of Ic being 1mA, Ohm's Law gives us

$$R2 = \frac{V}{I} = \frac{6}{0.0001} = 6000\Omega$$

Amplifier design

There is more than one way of connecting a transistor into a circuit that offer better temperature stability and gain. You need to remember that Figure 5.14 shows **a common emitter amplifier.**

You should be cautious when selecting transistors as the gain of a transistor, β, can have a wide variation in manufacturing tolerance.

Transistor oscillators

If the amplifier output is connected back to the input in some way, this can cause an oscillation, as anyone who has ever held a microphone too close to a loudspeaker can tell you! If the feedback path is frequency selective, then the frequency of oscillation can be controlled. Quartz crystals or combinations of inductors and capacitors can be used to achieve this frequency selectivity, and to create oscillators producing RF frequencies. This is discussed in more detail in the RF Oscillators chapter which we come to next.

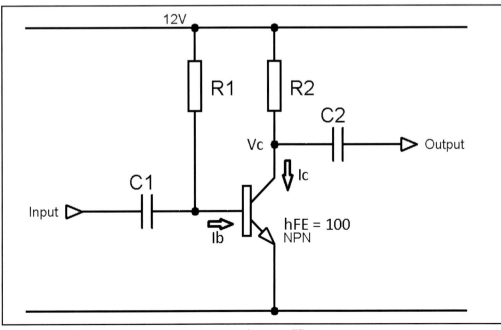

Fig 5.14: A simple single transistor common emitter amplifier

5: Semiconductors

6: RF Oscillators

Oscillators

An electronic oscillator is a device that produces a sine wave. Electronic Oscillators convert direct current to an alternating current and they are extremely important in radio design to produce audio and radio frequencies.

You will be familiar with the howling noise made when an amplifier microphone is placed too close to the loudspeaker. The sound, which is a vibration of the air molecules made by the movement of the cone in the speaker, is picked up by the microphone. It is amplified again and the cycle of *feedback* continues to produce the characteristic wailing note.

Essentially, oscillators are amplifiers where the amplified signal is fed back into the amplification circuit. For radio circuits we do this in a very controlled way. The amplified signal is electronically fed back into the amplifier circuit and the components we use ensure we get an output frequency that is *stable* and does *not* drift.

We are interested in audio and radio frequencies,
- Radio frequencies of about 100kHz to 1000GHz
- Audio frequencies of about 16Hz to 20 KHz

Although there are many differing designs of oscillators, we are going to concentrate on a circuit containing a single transistor amplifier.

Crystal Oscillator

There are three parts to the circuit,
1. a transistor acting as an amplifier.
2. a feedback from the transistor emitter to the junction between C1 and C2
3. a quartz crystal with trimmer capacitor. This is the frequency determining part of the circuit. The trimmer capacitor allows for small variations in frequency.

Crystal oscillators use a thin slab of Quartz as the frequency-determining element. Quartz is a natural mineral that exhibits a property known as the piezoelectric effect. It changes shape when a potential difference is applied to it. The crystal responds most strongly when

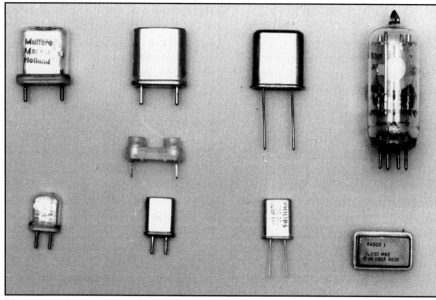

Picture 6.1: Examples of crystal oscillators and holders

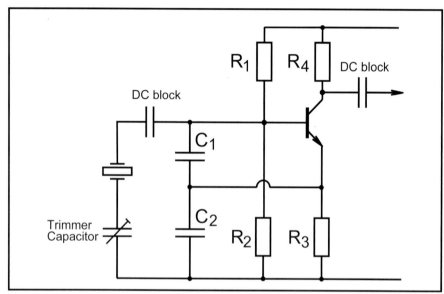

Fig 6.1: A basic Colpitts-type Crystal Oscillator

an AC signal at the crystal's resonant frequency is applied. (Physicists would say the crystal is being used as an electro-mechanical resonator.) By carefully cutting the quartz to a particular shape and thickness the resonant frequency can be precisely controlled.

Crystals are usually mounted in metal or less commonly glass enclosures with metal plates on either side of the thin quartz slab. As can be seen in Picture 6.1 they have either wire terminals that are suited for soldering into a circuit or stiff pins that enable them to be mounted into a socket.

Since the frequency of the crystal is fixed by its physical properties, it is not easily variable. Small changes in frequency can be achieved by changing or inserting a series reactance such as a small value capacitor or inductor. Crystal oscillators are used where you want a single precise, stable frequency. They are commonly used in filter circuits. At the full Licence level you will meet the *phase locked loop*, a variable frequency oscillator, which uses the crystal oscillator as a known frequency reference.

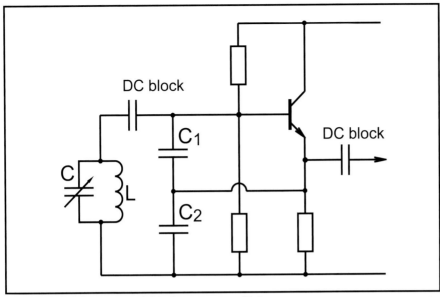

Fig 6.2: A Colpitts-type variable frequency oscillator

Variable Frequency Oscillator

The crystal oscillator is limited in its use in that it can only produce one frequency. We can of course have a circuit with switched crystal oscillators but this does not give us a continuous range of frequencies. *A variable frequency oscillator* (VFO) can be made by using the above circuit but replacing the crystal and trimmer capacitor with a tuneable Inductance/Capacitor (LC) stage. This is known as a Colpitts Oscillator (Figure 6.2) and named after its designer (Edwin H Colpitts in 1918).

The output frequency can be adjusted by varying either the value of the Capacitor, C, or the Inductance, L. The circuit shown uses a fixed value inductance and a variable capacitor.

You will not be expected to remember the full circuit diagrams for the exam, but you will need to be able to identify a crystal or LC oscillator. The frequency range of the VFO will depend on the maximum and minimum values of the variable capacitor (or variable inductor) used in the circuit. A capacitor with a larger capacitance range will give a wider frequency tuning range. While the VFO may seem advantageous over the crystal oscillator due to its tuning range, it does suffer some disadvantages. Firstly, because the oscillator is built to allow its frequency to vary, it will tend to drift and wander from the frequency it has been set to. Should this happen you may be transmitting outside the amateur band in violation of your licence. This can be minimised by:

- Providing the oscillator with a clean, well-regulated power supply.
- Using careful construction techniques, such as rigid mounting of parts.
- Use of screened enclosures help stabilise the temperature and keep stray RF from interfering with the circuit.
- Following the oscillator with a good buffer circuit to isolate the oscillator from the rest of the radio circuits.
- Using temperature-stable components.

It is critical a VFO is stable and drift free and calibrated to at least show the amateur band edges. Full calibration is a good idea so that you know your exact operating frequency and you do not cause interference to other radio spectrum users.

Audio Frequencies

Theoretically we can use the Colpitts VFO to generate audio frequencies. This is very difficult in practice as the values of L and C are very high. Suffice to say we still use a transistor as an amplifier but the method of electronic feedback is a little more complex and you are not required to know any of the details.

Direct digital synthesis (DDS)

Direct digital synthesis (DDS) is a technique that uses a mixture of digital and analogue electronics to produce a sine-wave at frequencies from audio to RF. Both the digital and analogue electronics are combined into a single integrated circuit. The IC contains a 'lookup table' – a table of pre-set digital values in memory – that represent the shape of a sine wave. The output of the lookup table is connected to a digital to analogue converter (DAC) which generates an analogue voltage from the digital value. The more 'bits' the lookup table contains and the DAC can process, the better the precision of the analogue output waveform and the fewer spurious emissions the DDS creates. DACs and their properties are discussed in more detail in the Transmitters and Receivers chapter, later on in the book.

On the input to the lookup table is a digital counter. This counter controls how fast the numbers in the lookup table are output to the DAC, and thus directly controls the frequency of the analogue sine wave at the output.

Low end DDS ICs are available very cheaply and are extremely commonly used where digital control or generation of an analogue waveform is required. To remove some of the spurious signals the DDS outputs, they are often followed by filtering. DDS principles are commonly used in conjunction with other techniques such as phase-locked loops.

Most modern oscillators are digital synthesizers because they are very stable. Their output is continually controlled by constant reference to the frequency produced by a crystal oscillator.

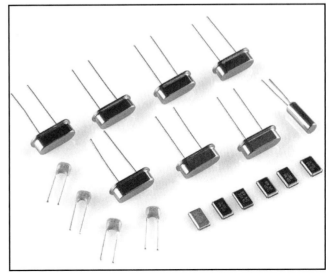

7: Transmitters & Receivers

Transmitters

A simple transmitter such as we saw in the Foundation licence is composed of several building blocks:
- An audio stage, that take signals from a microphone
- An oscillator that produces a radio frequency signal
- The modulator that combines the audio signal and the radio frequency signals
- A power amplifier that raises the modulated signal to a sufficient strength to transmit.

We will now look at how the following types of transmitters operate:
- CW (Morse Code)
- AM (amplitude modulation)
- FM (frequency modulation)
- SSB (single side band)

We will start by looking in greater detail at the common blocks shared between all of these transmitters, and then consider the blocks specific to each transmitter type.

Microphone amplifiers

The electric signal generated by a microphone is very weak, perhaps just a few milli-Volts. The signal needs to be amplified to a level sufficient to feed to the modulator. Often the microphone amplifier stage will filter the audio to limit it to a bandwidth of 300Hz to 3kHz – a typical bandwidth used for voice communication. The amplifier will be constructed from either discrete transistors or an integrated circuit.

RF Power Amplifiers

An RF power amplifier takes the low-level signals from the other modules in the transmitter and raises the power to a level suitable for transmitting. A practical power amplifier module will contain several transistors of either the bipolar or the FET type. You will not be expected to know actual circuits for the exam.

The RF power amplifier is almost always followed by a low-pass filter to reduce harmonics in the radiated transmit signal which can cause interference to other amateurs and radio users.

Mixers

To produce the transmitted signal, the audio signal between 300Hz and 3kHz must be converted to a radio frequency signal in the amateur bands; this conversion is performed by the modulator – a device that mixes the audio with the radio frequency produced by the local oscillator.

Depending on the type of mixer used, the frequency output of the mixer may contain some or all of the following:
- Either of the two input signals
- The sum of the two input signals
- The difference between the two input signals
- Any harmonics of any of the above frequencies

Typically, only some of these signals are of interest, and so the undesired signals are usually filtered out.

Example

Audio from a microphone is amplified and filtered to fit the range 300Hz to 3kHz. It is mixed with a 51MHz local oscillator. What frequencies are present at the output of the mixer?

Sum: 51MHz + 300Hz to 3kHz = 51,000,300 to 51,003,000Hz

Difference: 51MHz – 300Hz to 3kHz = 50,999,700 to 50,997,000Hz

In a transmitter, the mixer is often called a modulator and the mixing process is called modulation. In the case of a receiver, the mixer is often called a demodulator or a detector and the process is called demodulation.

Filters

Low pass filters

The last circuit in a transmitter before the antenna connection should be a low pass filter. The filter prevents harmonics of the wanted signal (the fundamental), from being radiated. Harmonics are signals related to the fundamental signal, but higher in frequency. Often, these are odd multiples of the intended frequency (3x, 5x 7x, and so on), but even harmonics may also be present. Harmonics can be generated by any part of the transmitter but are particularly likely to be produced by the power amplifier.

A low pass filter should be designed to

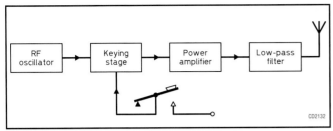

Fig 7.1: Block diagram of a CW Transmitter

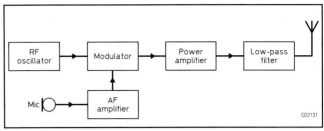

Fig 7.2: Block diagram of an AM Transmitter

cut off signals just above the fundamental frequency. For example, an 80m transmitter operating on 3.6MHz could have a filter that cuts off all signals above 4MHz, preventing harmonics from being radiated on 7.2, 10.8 and 14.4MHz (all just outside the amateur bands 40m, 30m and 20m).

Band pass filters

A band pass filter cuts off signals above and below its design frequency. The frequency the filter has been designed to pass is called the centre frequency. The filter will also allow frequencies in a range slightly above and slightly below the centre frequencies to pass through – this is called the filter's pass-band.

A good transmitter design will use a combination of band-pass and low pass filters to ensure the cleanest possible (lowest harmonic content) radiated signal.

Buffer Amplifiers

It is a bad idea to directly connect the output of an oscillator to any other stage without using a buffer amplifier. The purpose of the buffer is twofold; firstly, it serves to isolate the oscillator from anything else in the transmitter that may affect the frequency stability of the oscillator. Secondly, the buffer amplifier may be used to raise the oscillator signal to a suitable level to drive a modulator or power amplifier. A buffer amplifier may be followed by a band-pass filter to ensure any harmonics created by the buffer are not passed on further though the transmitter.

CW Transmitter

While a very basic CW transmitter can be created with a single transistor, a crystal and a Morse key cut from an old soft drink can, this is not best practice. Simple designs often use the key to switch the oscillator on and off, and do not have any buffering between the oscillator and the antenna. This leads to a 'chirpy' transmitted signal as the oscillator is switched on and off, causing its frequency to drift.

A good CW transmitter will have the oscillator running continuously and have a separate keying stage that interrupts the signal path to the power amplifier. This helps prevent chirp caused by keying the oscillator and key-clicks caused by keying the power amplifier. The term chirp comes from the characteristic sound of the defective CW signal at the receiver end, caused by the received tone frequency changing throughout the duration of each dot or dash.

Key clicks are caused by too fast a rise (or fall) time of the RF power output and are heard as clicking in time with the CW but audible on many frequencies right across the band. Key clicks can be reduced by a simple filter in the transmitter keying stage.

Picture 7.1: Morse keys come in a variety of shapes and sizes.

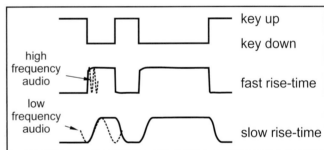

Fig 7.3: RF signal changes when key clicks are heard

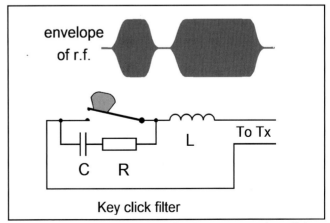

Fig 7.4: This is an example of a circuit to suppress key clicks

7: Transmitters & Receivers

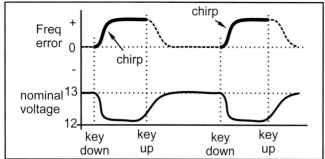

Fig 7.5: Chirp. Change in frequency is shown by the variation in voltage on the key *down* and *up* positions

You need to be able to recognise how the amplitude of the RF signal changes when Key Clicks and Chirp are heard.

Key clicks are nothing to do with the speed at which the operator is sending a CW signal. They arise because of the rise-time of the carrier when the key is depressed. A very fast rise time can be thought of as a small part of a high frequency audio signal and it produces very wide side bands. Conversely, a slower rise time equates to a lower frequency audio signal. A Key click filter is shown in **Figure 7.4**.

Chirp

Figure 7.5 shows how the transmitted frequency varies in the very short space of time after the key has been depressed. On key down the transmitter is drawing large amounts of current. To accommodate this, the ideal transmitter will have a well-regulated power supply producing a constant voltage. If this is not the case the variations in voltage will affect the oscillator causing changes in the frequency it produces.

AM Transmitter

The output of an AM transmitter is a signal of varying amplitude, which corresponds to the audio signal from the microphone. The output from the modulator contains two sidebands, one above and one below the carrier (fundamental) frequency. These sidebands contain the audio component of the signal. The carrier frequency and both sidebands are passed to the power amplifier and are transmitted. Recall that the usual bandwidth used for voice communication is 3kHz, in an AM signal, each sideband is 3kHz wide, so the bandwidth of a transmitted AM signal is 6kHz.

When the two signals are mixed, the levels of the RF carrier and audio signal control the depth of modulation. The depth of modulation describes how much the carrier amplitude changes for a change in audio signal amplitude and is defined as the ratio of the peak audio signal level to the unmodulated audio signal level, and is usually expressed as a percentage. With a modulation index of 50%, the peak amplitude of the modulated signal is 50% larger than that of the RF carrier alone, and the minimum amplitude is 50% smaller. Ideally, the modulation index is close to 100%, but it must never exceed 100% as this causes distortion and harmonics of the audio signal, which in turn will cause interference on adjacent frequencies – this is commonly referred to as 'splatter'. **Figure 7.6** shows how the depth of modulation of the signal changes.

SSB Transmitter

The single side band (SSB) transmitter is an evolutionary development of the AM transmitter. As the name suggests, an SSB transmitter only transmits one of the two sidebands and no carrier. This has the immediate advantage of halving the transmitted bandwidth – a useful effect on today's crowded bands. Instead of occupying 6kHz as with an AM signal, an SSB signal only occupies 3kHz since only one of the two sidebands is transmitted.

By filtering out the carrier frequency and only selecting one sideband for passing to the power amplifier, the SSB transmitter can concentrate its power into just the part of the signal that contains information – a carrier signal contains nothing useful, and the other sideband is just a copy of the first sideband with no extra information. This makes SSB more efficient than AM (and FM, as we shall see) in terms of transmitter power and bandwidth use.

The SSB transmitter includes two new stages we have not yet seen; these are the **balanced modulator**, and the **sideband filter**.

The balanced modulator is like the modulator used in the AM transmitter, it mixes the audio from the microphone with the RF from the local oscillator, but crucially it **only produces the sidebands**

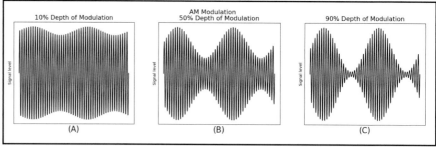

Fig 7.6: Effect of changing depth of modulation in AM modulated signal

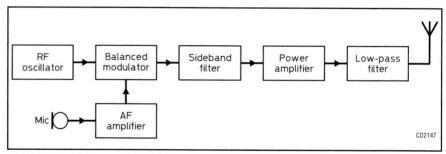

Fig 7.7: Block Diagram of an SSB transmitter

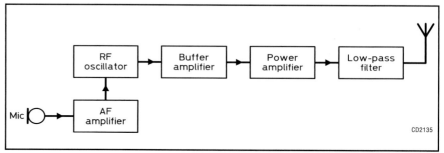

Fig 7.8: FM Transmitter Block Diagram

without the carrier, the original RF carrier frequency is suppressed.

The sideband filter is a band-pass filter with a very narrow bandwidth; its purpose is to only allow one of the two sidebands to reach the power amplifier, with the other sideband removed by the filter. Sometimes the sideband filter will be switchable, so that either the upper or lower sideband can be selected. This is common on multi-band transmitters for the HF bands, where lower side band (LSB) and upper side band (USB) modulation are used on different bands. Such a filter is typically made from several quartz crystals or similar resonators.

It is helpful to refer to **Figure 4.38** and re-examine the characteristics of the band pass filter. In the example shown we can see that the filter passes a relatively narrow range of frequencies at the peak of the curve but because of the wide skirt a significant amount of signal is also passed over a much wider frequency range. It would not be very selective. A sideband filter has to be highly selective. The upper and lower side bands are symmetrical about the carrier frequency (which has been removed) and extend from 300Hz to 3kHz into the upper and lower sidebands. This means that the filter must reject a signal only 600Hz away. Consequently, the sides of the response curve must be as steep as possible. This is achieved by using a crystal filter. It is common to use several crystal filters in a ladder configuration. You should be able to recognise and know the application of this arrangement. (**Figure 7.9**)

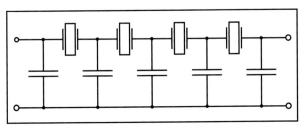

Fig 7.9: A crystal sideband filter; by using several crystals in a ladder configuration a suitably narrow pass band is obtained.

Type of Transmission	Bandwidth
Frequency Modulated, FM	Around 10kHz
Amplitude Modulated, AM	6kHz
Single Side Band, SSB	3kHz
Carrier Wave, CW	100 to 200Hz
Digital signals	May use less bandwidth than any of the above.

Table 1: Types of Transmission and their Bandwidth

FM Transmitter

Unlike the previous transmitters, the FM transmitter maintains constant output amplitude, but instead turns the audio from the microphone into a varying output frequency, commonly referred to as deviation. The amount of frequency deviation is proportional to the signal amplitude (how loud you speak into the microphone, for example) and not frequency of the signal.

This frequency deviation is commonly achieved using a special type of diode that varies its capacitance in response to an electrical signal: a variable capacitance or 'varicap' diode. This varying capacitance connected to the oscillator causes a corresponding varying frequency output. You may hear terms such as wide-band or narrow-band frequency modulation; these correspond to the amount of frequency variation permitted.

Its symbol is shown in **Figure 7.10** and like all diodes it has a depletion layer. We can exploit an unusual property of the depletion layer. It acts like the plates of a capacitor. By increasing the reverse bias the depletion layer gets larger and the capacitance decreases. Hence, by varying the amount of reverse bias we can control its capacitance.

The frequency deviation from the nominal carrier frequency is proportional to the amplitude of the modulating signal – the louder you speak into the microphone, the larger the frequency deviation. Unlike with AM, there is no natural limit to the amount of deviation so limits are imposed based on the purpose of the transmission and quality required. The **peak deviation** refers to the maximum frequency movement permitted from the nominal carrier frequency in either direction.

FM broadcast radio uses a peak deviation of ±75kHz which gives good quality sounding audio but is 'expensive' in terms of bandwidth required. Amateur transmissions use a peak deviation of either ±2.5kHz (12.5kHz channel spacing) or ±5kHz (25kHz channel spacing) which uses far less bandwidth while maintaining an acceptable audio quality.

The **modulation index** of an FM transmitter is defined as the ratio between the peak deviation and the maximum audio frequency.

$$h = \frac{\text{peak deviation}}{\text{maximum audio frequency}}$$

The maximum audio frequency used in an FM broadcast transmission is typically quoted as 15kHz. The modulation index (h) of an FM broadcast transmission is therefore:

$$h = \frac{75\text{kHz}}{15\text{kHz}} = 5$$

This is considered to be **wideband FM** since the modulation index (h) is greater than 1.

For an amateur transmission on the 2-metre band using 12.5kHz channel spacing, the modulation index (h) would be:

$$h = \frac{2.5\text{kHz}}{3\text{kHz}} \approx 0.83$$

This is considered to be **narrowband FM** since the modulation index (h) is less than 1.

Bandwidth and Choice of Transmission Mode

The bandwidth of a signal is the total range of frequency required for a modulated signal to be transmitted without loss of data or to become distorted. This is a significant point.

At Intermediate Level you are only required to know the relative bandwidths of the different modes but as you are now a practising amateur it helps to be aware of some of the background theory to help develop your interest in the hobby.

FM signals, in theory, have an infinite number of sidebands which extend out further than the peak deviation. Carson's Rule, which you will meet in the Full Licence gives us a simple formula for calculating FM bandwidth,

Bandwidth = 2 x (maximum audio frequency + peak deviation)

A VHF signal with a maximum audio frequency of 2.8kHz and a peak deviation of 2.5kHz, will have a Bandwidth = 2 x (2.8 + 2.5) = 10.6kHz

Fig 7.10: Symbol for a variable capacitance diode

AM and SSB

The information required for adequate understanding of the human voice is carried in the frequency range of 300Hz to 3kHz. It follows that an AM signal (containing both the lower and upper sidebands) has a bandwidth of 6kHz. By removing one of the side bands as in Single Side Band transmission the bandwidth is reduced to 3kHz.

Carrier Wave.

CW, has a much smaller bandwidth. It is not related to the speed with which the operator sends the message; it is more a function of the rise and fall of the signal with the strokes of the key and the way in which the signal is smoothed. This is fully dealt with in the Full Licence.

Digital modes

Digital modes, of which there are many examples, have a very much smaller bandwidth. A typed message is prepared on a computer keyboard. A digital signal is produced which is fed into a sound card to make audible tones. These are forwarded to the microphone socket of the transceiver. If the audible tones are too large there will be distortion and an excessive bandwidth will follow. It is important to configure the computer correctly. This is very similar to ensuring the correct microphone gain for an SSB transmission.

The general acceptance that digital modes have a very narrow bandwidth needs careful consideration. This is not always the case and will depend on the rate at which data is being sent and the nature of the mode being used.

Harmonics; Removal by Filtration

Harmonics are best explained with an example. Let us assume we are working with a wave of 100Hz frequency. This frequency is known as the *fundamental frequency*. The second harmonic will have a frequency of 2 x 100 = 200Hz, the third harmonic will have a frequency of 3 x 100 = 300Hz and so. Be careful with this definition as it is a common mistake to wrongly say that the first harmonic would be 200Hz. There is no such thing as a *first harmonic*.

Transmitters contain oscillators, mixers and amplifiers which produce harmonics, multiples of a fundamental frequency, and if they are not filtered out can cause interference in other amateur bands and to other radio users.

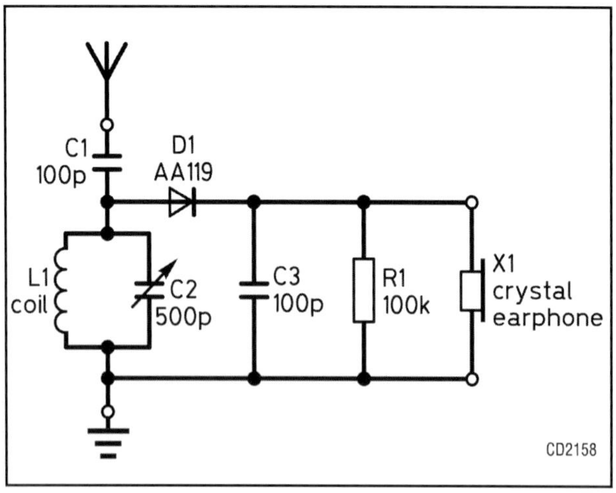

Efficiency of RF Power Amplifier

There are three main classes of amplifier we can use for RF power and they will be fully specified in the Full Licence.

We can estimate the amount of RF power leaving the RF amplifier. Let us assume the DC supply is 13.4 Volts which is a typical value for a small portable unit and a current of 15 Amps is drawn by the RF amplifier upon transmission.

Classes of Amplifier	
Type	Efficiency %
Class A	35
Class B	50
Class C	67

Power (Watts) = Amps x Volts

Power supplied to amplifier =
13.4 Volts x 15 Amps = 201 Watts

If we are using a Class A amplifier which has an efficiency of 35%,

the power leaving the amplifier

= 0.35 x 201 Watts

= 70.35 Watts.

Receivers

The Intermediate licence builds upon your knowledge of radio receivers from your Foundation studies. We shall look at the building blocks of receivers in some detail and see how they are used in common receiver designs.

The blocks encountered in the Foundation licence are:

- A tuning stage to select the wanted frequency from all those present at the antenna.
- A detector that turns the radio frequency signal into an audio signal.
- The audio amplifier that increases the level of the detected audio signal to drive a speaker.

The Crystal Set (or Diode Receiver)

This was the earliest type of receiver invented, and they can still be used to listen to AM radio transmissions today.

The crystal set (**Figure 7.11**) starts with a tuned stage consisting of an inductor and capacitor that form a parallel tuned circuit with a high impedance at the frequency to be received. All other frequencies pass through the circuit to ground and as the received frequency cannot pass through to ground it is passed to the detector.

The detector in a modern crystal set is a germanium diode because their low forward voltage, allows them to be sensitive to very low-level signals. The diode demodulates the AM RF signal producing

Fig 7.11: circuit diagram of a crystal-set radio receiver. If you wish to build one you can use this circuit as a guide, the coil L should be around 80 turns of wire on the cardboard tube

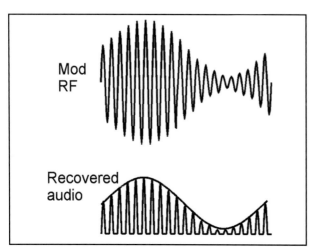

Fig 7.12: The diode demodulates the RF wave

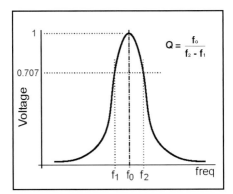

Fig 7.13: Q-factor voltage curve

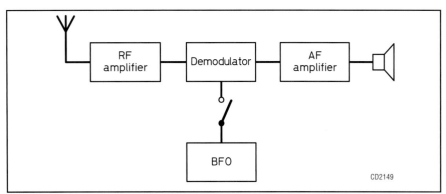

Fig 7.14: Block diagram of the TRF receiver.

a varying voltage that can drive a high impedance earphone. The audio levels are low, because there is no amplification; the whole circuit is powered just by the energy picked-up by the antenna.

The diode receiver is rather limited, it can only receive strong signals and will quite often receive several signals at once – it exhibits poor sensitivity and selectivity.

You do need to know how the diode demodulates the RF wave. As the RF signal passes though the diode it is rectified and the negative part of the wave is removed. Around the remaining part of the wave we can see a smooth envelope. This envelope is a voltage shifted representation of the original audio signal. The signal passes into the headphones and the audio is recovered.

Receiver Sensitivity and Selectivity

The words sensitivity and selectivity can be a little confusing to the new comer to the hobby. In short sensitivity is a receiver's ability to detect weak signals. At the intermediate level this is very subjective. You can either hear a signal or you cannot. In the Full Licence this is explored further and we can express a receiver's sensitivity with numbers and units. You need to be aware that very strong signals can overload a receiver and distort the audio output.

Selectivity is the ability of a receiver to reject signals outside the wanted band width. Obviously a good receiver will select only the frequency we want but as this is a scientific hobby we need a more rigorous understanding.

The diagram in **Figure 7.13**, shows how the amount of signal selected varies with frequency. The signal passed increases to a maximum and drops as we approach and pass f_o. The shape of the curve tells us something very important; the wider it is the greater the amount of another signal of frequency near to f_o will pass into the receiver. Said another way; a wide curve is not very good for selectivity.

By convention we say that 100% of the signal is passed at f_o, (0dB). On either side of f_o, there is a lower and higher frequency, f_1 and f_2, where the signal strength is 50% (-3dB) less than the maximum. We call the frequency range between these two points the band width or more correctly the half power band width.

(Notice that the power has been represented as a voltage. Power is proportional to V^2. You will find it beneficial to satisfy yourself that at half power the voltage is reduced to 0.707 of its original value)

The breadth of the curve can now be defined by its Q or quality factor.

$$Q = f_o / \text{bandwidth}$$

Hence,

$$Q = \frac{f_o}{f_2 - f_1}$$

Q has no units; it is a simple number and it gives us a measure of how selective the tuned circuit is. Q values of up to about 70 are reasonably straightforward to achieve, but over 100 is difficult.

As you progress into the hobby you will see that Q is sometimes referred to as the voltage magnification in a series resonant circuit and the current magnification in a parallel resonant circuit. This will be explored more fully in the Full Licence but it is something you should be aware of as there are important safety concerns when working with tuned circuits. You will learn two things;

In a series resonant circuit the voltages across the Inductor and Capacitor are not only equal but much higher than the voltage of the applied or external circuit.

In a parallel resonant circuit the current in the Inductor and Capacitor are not only equal but much higher than the current in the applied or external circuit.

$$Q_{Series\ Circuit} = \frac{Voltage_{Across\ L\ or\ C}}{Voltage_{Applied}}$$

$$Q_{Paralell\ Circuit} = \frac{Current_{In\ the\ Inductor\ or\ Capacitor}}{Current_{In\ external\ Circuit}}$$

Q can have values of 100 and this means that the voltage in a series resonant circuit and current in a parallel resonant circuit will be one hundred times greater than that of the external circuit and present a significant danger in transmitter circuits.

The frequency pass band of a receiver is specified so as to not to cause significant attenuation of the wanted signal. It may be as narrow as 100Hz for CW and as wide as 2.5 to 3kHz for SSB. Receiver design is aimed at getting rejection of unwanted frequencies at the -60dB (one millionth) power level.

Tuned RF receiver

The Tuned Radio Frequency receiver is not a syllabus item and is not examined. It does however contain important learning points which will help in your understanding of other parts of the syllabus. The tuned radio frequency receiver (TRF) is very similar to the crystal set or diode receiver, but with a couple of important changes. You can see from the diagram (**Figure 7.14**) the first block connected to the antenna is an amplifier. This is a tuned amplifier, it only amplifies a narrow band of frequencies around the signal you wish to receive, other signals are blocked or not amplified; this improves the selectivity of the receiver. A weak signal you wish to listen to doesn't become swamped by a strong local signal as easily as can happen in a crystal set.

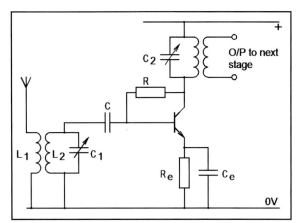

Fig 7.15: Circuit diagram of the TRF receiver input stage

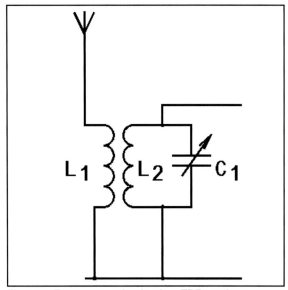

Fig 7.16: Frequency selection in a TRF receiver

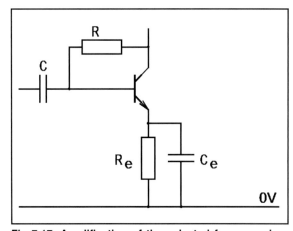

Fig 7.17: Amplification of the selected frequency in a TRF receiver

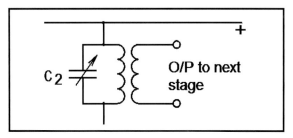

Fig 7.18: A tuned circuit and transformer coupling to the demodulator

The demodulator is often a single diode as is used in a crystal set.

A TRF will also commonly have an audio amplifier at the output so a speaker or headphones can be used at a comfortable volume.

You may notice the block marked BFO in the diagram; this stands for beat frequency oscillator and is not present on all TRF receivers. It is used to enable reception of CW and SSB signals. The BFO is explained in subsequent sections.

At Foundation Level you will have seen schematics for receivers and transmitters which have been broken down into block diagrams. Each block represented a circuit designed to carry out a specific task. Although this is a good ploy to gain an overall understanding of what is happening the Intermediate Licence teaches you how individual components work together and you need to be able to recognise the configurations of components and their combined purpose.

There are two general points which apply to all circuit diagrams.
• A DC power supply is indicated by the lines at the top and bottom of the circuit labelled + and 0V. Sometimes the bottom line is labelled –ve (remember we are using conventional current.)
• Radio designers have a general rule of showing how a signal is processed as it passes from the left hand to the right hand side of the main circuit.

We can separate the circuit shown in **Figure 7.15** into three smaller parts,

Using a transformer in **Figure 7.16** the signal from the antenna is passed into the tuned circuit. With the variable capacitor we select the frequency we want.

Picture 7.2: A double ganged variable capacitor

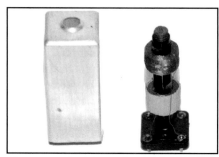

Picture 7.3: Inside an IF transformer.

The output from the tuned circuit passes into the base of a transistor shown in **Figure 7.17** for amplification. Note that there is a capacitor before the **base** of the transistor to pass the AC signal but block any DC. The transistor amplifies only the resonant frequency and all other frequencies are grounded.

As all other frequencies have been grounded the selected frequency can be transformer coupled to the next stage, **Figure 7.18**, in the receiver for demodulation.

As can be seen in **Picture 7.2** a common feature of TRF receivers is for the variable capacitors in the tuning and output stages to be a double ganged variable capacitor where the two sets of vanes move together when the spindle is turned. This allows the tuned circuit with C1 to be set to the same frequency as the circuit with C2. See also **Figure 7.19**

Superheterodyne receiver

The TRF receiver is a large improvement over the simple crystal radio, but does not offer enough selectivity (the ability to discriminate between multiple stations closely spaced in frequency) for the busy amateur and commercial bands today. The superheterodyne (or 'superhet') receiver (**Figure 7.20**) is the commonly used solution to this problem. The superhet is the most popular form of analogue receiver in use today.

From the block diagram you can see we add several new blocks to the simpler TRF type receiver. Taking them in order from the antenna we have:

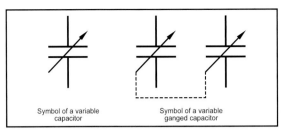

Figure 7.19: Ganged variable capacitors.

- A mixer connected to a Local Oscillator. The mixer mixes (or 'heterodynes') the incoming signal with the local oscillator to produce an intermediate frequency (IF)
- The IF amplifier in **Figure 7.21** is a fixed frequency amplifier with narrow filtering; it produces gain in the radio at a single frequency, improving the sensitivity of the receiver. Often the IF amplifier block will be composed of several different amplifier circuits in series.
- Demodulator (sometimes detector) and BFO: The demodulator can be as simple as a single diode as used in the TRF and crystal set, but will usually be more complex in a superhet. Typically, the IF signal is mixed down to audio using the BFO signal, allowing a wider range of modes to be demodulated.
- Audio Amplifier: These can be as simple or as complex as required. Due to the gain produced by the IF amplifier, the audio amplifier will not have to amplify as much as in a TRF, this allows the designer some freedom to construct a pleasant-sounding amplifier.

You need to be able to recognise an IF amplifier circuit as is shown in Figure 7.20. It is very similar to that used for the TRF

The Intermediate Frequency produced by mixing is fed into the transistor amplifier circuit. The tuned circuit ensures that the intermediate frequency has a bandwidth to allow just the wanted signal and reject others on adjacent frequencies. Frequencies just above and below the IF are passed to ground and the output to the demodulator is via the IF transformer.

Important points to note,
- The primary and secondary coils are wound on a plastic former.
- Through the centre of the former is a thread supporting an adjustable iron or ferrite slug.
- The transformer is enclosed in an aluminium can to provide RF protection.
- Sometimes a small capacitor called a trimmer capacitor may be enclosed in the can (or sited nearby and shielded)
- Fine tuning to obtain the required Intermediate Frequency is obtained by adjusting the trimmer capacitor and position of the slug.

Intermediate frequencies

Interference from signals close in frequency to the one you wish to receive can make it difficult for a simple receiver to recover just the signal of interest. You could in theory produce a very narrow band-pass filter to sit between your antenna and radio, but this would be difficult to construct and only work on a single frequency. Some very simple CW receiver kits available online do exactly this, at the expense of limiting their use to a single spot frequency on one band.

A superheterodyne receiver works by converting the frequency we wish to receive to an intermediate frequency (IF) where a good narrow filter can be built that will filter out unwanted signals. Recall that when you mix two different frequencies, you get both the sum and the difference of the two frequencies. By mixing the received signals from the antenna with a carefully chosen local oscillator frequency, it can be arranged that the signal of interest passes through the narrow IF filter. The other mixing products are rejected by the filter.

If the local oscillator is a variable frequency oscillator, the reception frequency of the radio can be adjusted over a wide range, while maintaining the performance enhancement of the quality IF filter.

Example
It is common to build receivers for the 20m band using a 9MHz IF and a local oscillator tuning between 5 and 5.5MHz.

Let's see how this works to receive 14 to 14.5MHz:

$$14MHz - 5MHz = 9MHz$$

$$14.50MHz - 5.5MHz = 9MHz$$

This also works for any frequency between 14 and 14.5MHz, because the LO can tune to any frequency between 5 and 5.5MHz. You can see in this way we are using the difference frequency produced by the mixer and discarding the sum frequency.

Beat Frequency Oscillator

The signal at the output of the IF amplifier will depend on the type of transmission being received. Up to this point all the receiver has done is amplify and frequency shift the signal. Turning the signal into something a human (or increasingly, a computer) can understand is the job of the detector or demodulator. A word on naming, usually the single diode as used in a crystal set or TRF receiver is called the detector, more complicated circuits for turning RF into audio are usually called demodulators.

Using a diode to demodulate the IF signal only works for amplitude modulation transmissions. The TRF and the superhet can both include a beat frequency oscillator (BFO) to enable CW reception, it may also be possible to tune in SSB using the BFO if it is carefully adjusted.

The BFO may be another tuneable

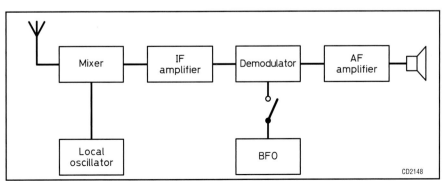

Fig 7.20: Block diagram of a super heterodyne (superhet) receiver

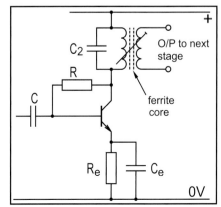

Fig 7.21: Intermediate frequency amplifier.

7: Transmitters & Receivers

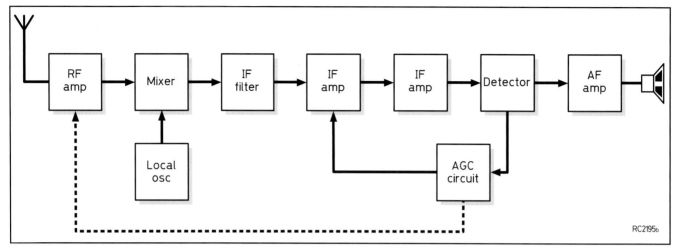

Fig 7.22: Block diagram of a superhet using AGC

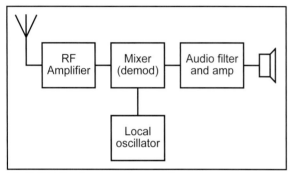

Fig 7.23: Direct conversion receiver

oscillator. It runs at almost the same frequency as the RF in the TRF receiver, or the IF in the superhet receiver. By carefully tuning the BFO to be about 1kHz away from the signal at the detector, the two signals mix to produce the sum and difference signals; one of which is at audio frequency and passed to the audio amplifier.

Example
A BFO at 9,001,000Hz will mix with an IF at 9,000,000Hz to produce a 1kHz difference. It will also produce an 18MHz sum frequency, but this is far outside the range of human hearing and would be rejected by the audio amplifier anyway.

To properly demodulate SSB signals, a superhet will use a demodulator called a product detector together with a fixed frequency oscillator known as a carrier insertion oscillator (CIO). The CIO runs slightly lower in frequency than the IF for USB and slightly higher in frequency than the IF for LSB, to mix with the SSB and reproduce the audio frequency signals. In effect, the CIO replaces the carrier that was removed by the balanced modulator in the transmitter.

FM signals require the use of a type of demodulator called a frequency discriminator to produce audio from the small changes in frequency at the IF. It is just about possible to tune FM signals on a receiver without an FM discriminator using a technique called slope-detection, but this is beyond what is necessary for this course.

Automatic gain control
Automatic gain control (AGC) is used in the majority of receivers to produce a constant audio level despite changes in the received signal strength. While there are many ways to achieve this, the more common way is to use signal strength at the demodulator and feed a control signal to the IF (and optionally RF) amplifiers (see **Figure 7.22**). If the signal strength increases, the feedback reduces the gain of the amplifiers, and when the signal strength decreases the amplifiers increase their gain.

This AGC control signal is often also responsible for driving the S-meter.

Direct conversion receiver
A direct conversion receiver does exactly what its name implies. In, essentially, in a single step it converts the RF signal into a demodulated audio signal. Let us assume that the frequency we want to demodulate (and listen to) is f_{RF}. By mixing this signal directly with another signal of the same frequency, f_{LO}, (subtractively) which is produced in the Local Oscillator we can directly recover the audio frequency. This pre-supposes the local oscillator being tunable over the range of frequencies you wish to receive. As the frequency output of the LO is varied a point will be reached where $f_{RF} = f_{LO}$ and demodulation occurs.

The frequency produced by addition in the mixer, $f_{RF} + f_{LO}$ is removed by the low pass filter. The direct conversion receiver has the advantage of high selectivity and is a precision demodulator

Be aware that it has some of the characteristics of a superheterdyne radio and it can be a little confusing when it is correctly referred to as a homodyne, synchrodyne or zero-IF receiver.

Receiver overview
It is possible to transmit both voice and data over AM, FM and SSB modes. CW is a special case where the carrier is periodically interrupted rather than modulated. These can all be demodulated using fairly simple circuits and techniques:

- AM with a single diode
- FM with a frequency discriminator
- CW with a single diode + BFO
- SSB with a product detector and Carrier Insertion Oscillator (CIO)

More complicated receivers can be built to optimise specific features, but these circuits are beyond the scope of the Intermediate licence.

Software Defined Radio
As you will have seen throughout this chapter, it is common to view a radio as a series of interconnected blocks, with each block performing a simple function on the signal passing through. Such examples include filtering, amplifying or demodulating signals. Conventional radios use many individual components to create these blocks. In a software defined radio (SDR), the functionality of each building block is instead implemented by computer software or digital logic performing

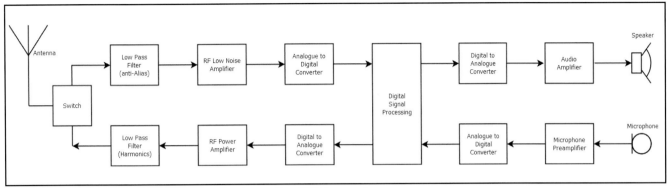

Fig 7.24: A simplified block diagram of a direct sampling SDR transceiver

the same basic tasks. Like conventional radios, software defined radios also come in several design variations. The simplest and most versatile is the direct sampling SDR. **Figure 7.24** shows a simple block diagram of such a transceiver.

Data Conversion

In an SDR receiver, the analogue to digital converter (ADC) block converts an analogue signal at the antenna socket, into a digital representation – a number – that represents the original input voltage at that precise time. Each time the ADC does this, it is said to have taken a sample. The frequency at which the ADC takes samples is referred to as the sample rate. This sets the maximum input frequency (or bandwidth) the receiver can handle; the ADC cannot process any input change between the samples. An ADC must sample at least twice as fast as the fastest input change to ensure that no change is missed – this minimum sample rate for a given input frequency is known as the **Nyquist rate**.

The accuracy of the samples is set by the ADC's resolution and is typically measured in bits, and describes the granularity (level of detail) of the samples. When a sample is made, the input voltage is converted and rounded to the nearest digital value the ADC can produce. Because the ADC rounds to the nearest level available, a small error is introduced, and the resulting digital representation is slightly distorted.

Sample rate and resolution bits are the basic figures of merit for ADCs. The faster sampling and/or higher resolution the ADC is, the more accurately the digital values will represent the analogue waveform. However, such an ADC is more expensive.

Digital to analogue converters (DACs), as the name implies, do the opposite to ADCs – they convert the samples of numeric values back into an analogue voltage. DACs have the same associated terminology of sample rate and resolution.

Signal Processing

The ADC sends each digital sample to the digital signal processing (DSP) block for processing. The incoming data represents the input signal in the time domain – that is, how the signal level varies with time. However, to be able to process the data to allow the user to select a frequency and demodulate the original audio, the SDR requires the signal to be in the frequency domain – this shows the level of every frequency present at the input.

A Fourier transform is a mathematical function which achieves this, computing the amplitudes of all constituent independent frequencies present in the incoming time domain data.

The power of the Fourier transform can be shown with a very simple example shown in **Figure 7.25**.

Examine the adjacent waves. If we assume that the time scale represents 1000ms or 1second we can see that,
1. **A** has three complete waves per second and its frequency = 3Hz,
2. **B** has two complete waves per second and its frequency = 2Hz
3. **C** has one complete wave per second and its frequency = 1Hz

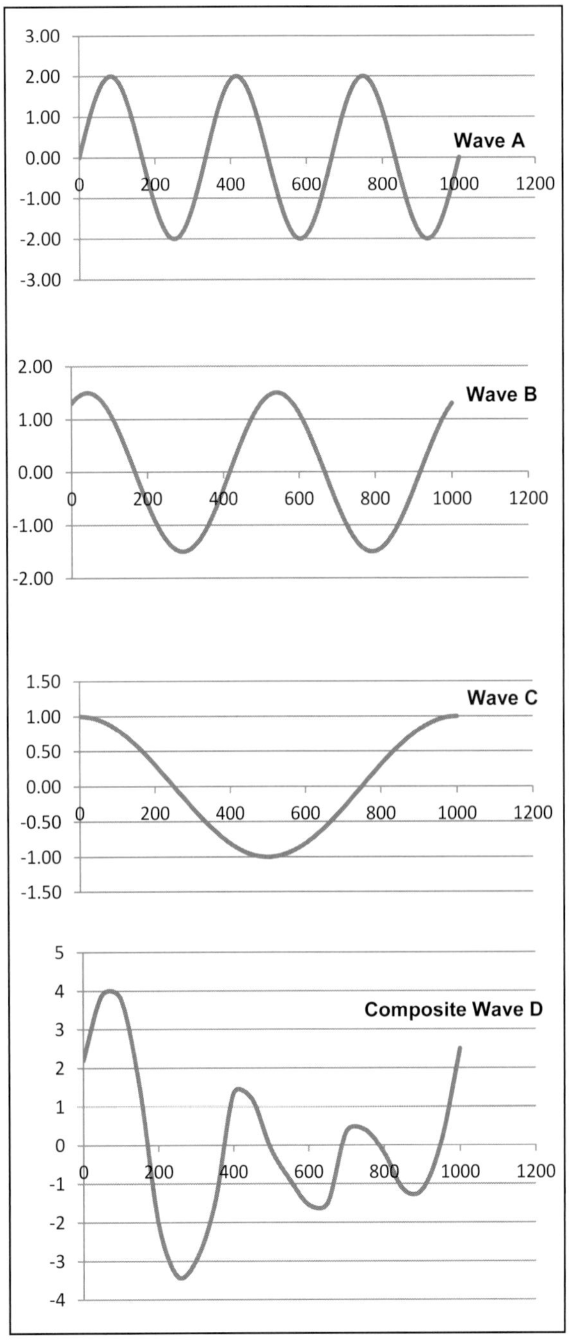

Signals B and A are the second and third harmonics of the fundamental, C. They

Fig 7.25: A composite signal formed from a single frequency and two of its harmonics

all have different amplitudes. In the real world we may have such a situation where

7: Transmitters & Receivers

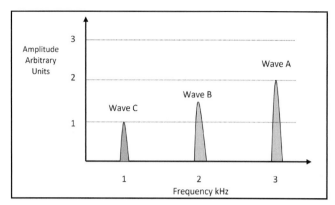

Fig 7.26: The separation of a signal into its fundamental frequency and two harmonics.

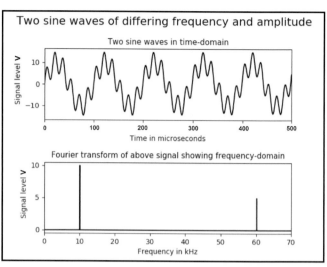

Fig 7.27: A Fourier transform can deconstruct a complex signal into component parts

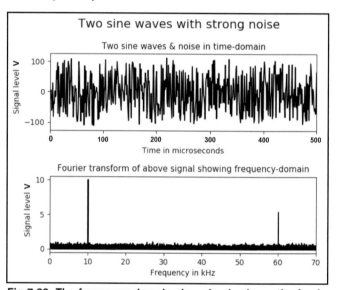

Fig 7.28: The frequency domain view clearly shows the fundamental frequency components when the time domain signal is lost into noise

harmonics are being generated and their combined effect would be shown by the Composite Wave as shown by the Composite Wave D.

The Fourier transform converts this time domain picture into the frequency domain as shown in **Figure 7.26**. It is impressive to see that the frequencies of the fundamental and harmonics have been clearly separated and their relative amplitudes displayed.

The output of the Fourier transform will show the amplitude of every frequency between DC (0Hz) and the highest frequency the system can handle, which is half of the Nyquist frequency. **Figure 7.27** aims to show this graphically.

The top section of Figure 7.27 shows a time domain view of two sine waves mixed together. It is possible to see that there are two independent waves, but it is quite difficult to measure their respective frequencies and amplitudes.

The bottom section of Figure 7.27 shows a Fourier transform of the signal seen in the top section. It should be much clearer to see what frequencies and their amplitudes are present: one sine wave has a frequency of 10 kHz and a peak amplitude of 10Volts, the other has a frequency of 60kHz and a peak amplitude of 5Volts. A Fourier transform provides a different view of the input data and is often used to give a spectrum or waterfall display such as shown in **Picture 7.4**.

Figure 7.28 shows how useful the frequency domain view can be when the presence of noise is considerably stronger than the signal. In the time domain, the two sinusoidal waves are totally obscured by noise, yet in the frequency domain, their presence is still very visible.

Digital Filtering

In a conventional transceiver the filter bandwidths are fixed by physical components: either expensive mechanical filters or combinations of inductors and capacitors. Typically, there are a few options; a wide filter for SSB and a narrow filter for CW or data. With an SDR's DSP (Digital Signal Processing) functionality, it is possible to implement any filter that is desired and since they are implemented in software there is no additional cost. Digital filters can be much more selective than analogue filters, as there are no resistive losses. The digital filter bandwidth can be made continuously adjustable, enabling the operator to carefully enclose a signal of interest, removing much of the background noise.

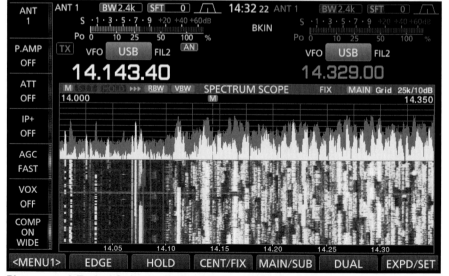

Picture 7.4: A Typical Software Defined Radio display

8: Good Radio Housekeeping

Good radio housekeeping is the practice of creating and maintaining a well organised, laid out and assembled shack. In doing so, you minimise the risk of causing interference to other electrical items such as your neighbours television, radio, telephone, computer, sound system as well as many more electronic devices that fill modern homes.

This section of the book is all about minimising your chances of causing interference.

Layout and Cables

It is good practice to keep RF cables away from all other cables to minimise interference. RF leads should use good quality plugs and sockets, with well screened coaxial cable between to minimise RF 'leaking' from the cable. Microphone and audio cables should be screened to help reduce their chances of picking up stray RF energy, and their screens should be properly connected to ensure consistent screening throughout their length.

Mains power cables also require consideration. If there is an indication that your transceiver may be causing interference you should check that it is not caused by RF leaving the transmitter via the power leads. A simple test for this would be to disconnect the antenna from the transceiver and reconnect the RF output socket to a Dummy Load and try a short transmission. The antenna will not be radiating a signal and the persistence of interference will indicate RF going into the power lead and hence the household supply.

Alternatively, you may suspect that you are receiving interference via the power lead. A simple way to prove this is to disconnect the power supply and operate the radio directly from a battery. This may not always be possible as many modern transceivers have integrated power supplies.

To prevent any RF from your transmitter entering your household mains supply, particularly the earth, it is a good idea to place a ferrite ring or clip-on ferrite around the lead as close to the equipment as possible. See **Pictures 8.1** and **8.2**. Ferrite Rings come in many different sizes and type of ferrite designed for particular tasks and frequencies. Using a ferrite also helps to reduce any mains-borne noise entering your equipment and deteriorating the receiver's sensitivity. In a mobile set up, allow space between cabling for your radio installation and the vehicles wiring loom to minimise the chances of interference. Stray radio frequency energy from a transmitter can cause interference to vehicle electronic circuits, such as your vehicle's audio and navigation systems, remote locking and alarm systems, and engine fuel management systems. The likelihood of occurrence is particularly increased when operating equipment with an RF output of 10W or more.

Even good quality transmitters from reputable manufacturers sometimes produce unwanted spurious signals and harmonic radiations. Fitting an external low pass filter will allow the wanted signal to pass and reduce the higher frequency harmonics as we have seen already. If you decide to do this, it should be inserted after the SWR meter and before your AMU.

Antennas

As you learnt in the Foundation training, it is strongly advised to keep the antennas as far away from houses as possible. It is even more important to do this if you decide to run higher power which you can do once you have gained your Intermediate Licence. This helps to reduce your chances of causing interference to your home and others, and, reduces the chances of your receiver performance being compromised due to the abundance of electrical noise from the modern home.

In cities and built up areas, there is a temptation to install your antenna in the loft of your home, but it is not advisable to do so. Wiring inside the loft will run all over the property, so any RF coupled into the power cabling is carried all around the house. Often the main TV reception antenna is in the loft along with any TV distribution amplifiers and so the potential for the RF being picked by the house equipment should not be under estimated. Installing a temporary antenna in the shack is also likely to cause you problems, since the antenna will create a strong RF field inside the room, likely to couple into any cabling around! Furthermore, there are obvious safety concerns being so close to a radiating antenna. When mounting antennas on vehicles, the antenna location will affect the radiation field strength within the vehicle. A wing or boot mounted antenna is likely to produce higher exposure levels inside the vehicle than a roof mounted antenna would.

Balanced antennas are less likely to cause interference, but it is worth repeating that antennas inside (or near) houses are best avoided where possible. If you are opting to use a vertical antenna, be aware that it will need a good earth connection. Without a good electrical earth (ideally, plenty of radials) there is a much greater risk of causing interference.

Earthing requirements

Your shack will require two different earths, each with a different purpose. Please take great care with these as confusing them could cause interference from your trans-

Picture 8.1: Ferrite rings on mains cables should be as close to the equipment as possible

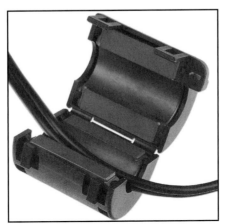

Picture 8.2: A clip-on ferrite

mitter to enter other equipment in your shack, but far more importantly, could cause a potentially fatal electric shock!

All electrical equipment should be earthed and your shack equipment is no different. The first earth we will talk about is the mains earth.

All mains powered equipment must be earthed for safety reasons. The only one exception to the rule is equipment that is 'double insulated'. Double insulated equipment is generally equipment that has no exposed metal surface but is positively identifiable by the double insulated symbol shown in **Figure 8.1**.

Fig 8.1: The double insulated symbol

As your radio equipment is unlikely to be 'double insulated' and will have exposed metalwork, it *will* need to be earthed using the earth pin in the 3-pin mains plug. The mains earth is for your safety and should **never** be removed.

The other type of earth is a RF earth which diverts RF current away from the mains earth and down to ground. The RF earth consists of one or more earth rods driven into the ground. Earth rods are made for this purpose and can be purchased from amateur antenna suppliers or a good electrical supplier dealing in domestic and trade sales. The rods are usually made from copper-coated steel and are about two metres long. You should hammer them into the ground close to your shack, making sure to avoid any underground cables, pipes and drains and building foundations. Then clamp a heavy gauge cable to the rod and connect directly to the transmitter or antenna matching unit, as shown in **Figure 8.2**. Sometimes it is helpful to run radials out from these ground rods, too.

Log keeping and complaints

Although it is no longer a condition of your licence to keep a log of transmissions made, keeping a log of all your contacts is part of good radio housekeeping. Amateurs are the only radio users allowed to design and build their own equipment. All other users must buy properly tested and approved transmitters. One benefit to radio amateurs is the allowance to transmit high powers in residential areas, with all the possibilities for interference that this brings, as we have seen above. That really is quite a privilege and why training and exams are necessary.

The logbook is very useful if an interference problem arises. Were you transmitting at the time of a problem? If not, you can quite quickly show you were not the source of interference problems. Your log will show you exactly what were you doing at a given time. Such information will help greatly in recreating the same situation, when investigating what the cause of the problem is.

If the person making the complaint is also willing to keep a log, it will help in identifying what transmissions are causing difficulties, since the two logs can be compared. If the matter becomes a formal complaint to the authorities, i.e. Ofcom, they will require logs to be kept as a first action in identifying the cause.

If you do receive a complaint of any kind, an honest, open and professional approach is always the best policy. Ask the person making the complaint to co-operate with the tests to identify the problem. Be as tactful as you can. You need to understand that they do not want the interference and they need to understand that you want to be able to transmit and enjoy your hobby. Show you are serious about resolving a case of interference, but also serious about being able to pursue your hobby.

Offer not to transmit at key times until a cure is found, but do not admit blame and do not make the offer sound like a permanent change. If the problem is caused by inadequate immunity in neighbours' equipment, as it often is, joint tests may demonstrate that without the difficulty of having to say as much. It is always useful to be able to show that your own TV and radio equipment is free from interference.

Seeking help with problem cases

If you have exhausted all your potential solutions but are still having problems, you may need assistance. The best source is the RSGB's EMC committee. The EMC committee produce several leaflets on EMC and interference. Some are aimed at the amateur, whilst others are intended for the person making the complaint. They all aim to provide unbiased and accurate information on interference matters. These leaflets are available by post or can be downloaded from the EMC Committee website (**Figure 8.3**), and contain tips that may help you solve a problem. If not, RSGB members may contact the EMC committee chairman who will provide the address of the nearest committee member who can assist.

Further help is available if these steps do not resolve the problem. Ofcom has useful advice on its website www.ofcom.org.uk but this will inform you that, the BBC is now responsible for investigating complaints of interference to domestic radio and television. Ultimately, if the complainer wishes to consult Ofcom their website describes the complaints procedure and how a complaint is registered.

Electromagnetic Compatibility

So far we have looked at how the RF we radiate from our equipment can cause problems with other radios and televisions. This is a small part of an important topic called electromagnetic compatibility (EMC).

We need to appreciate that all electronic equipment is capable of radiating and absorbing radio frequency energy. Most amateurs working in a built up area will complain of a very "high noise floor". The total effect of all the RF being radiated from domestic and commercial apparatus gives a considerable hiss that makes listening to weak signals difficult.

As the world becomes more reliant on

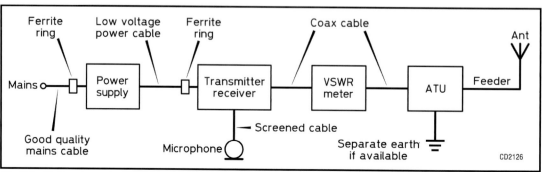

Fig 8.2: Station layout with filters on power leads and correct earth connection

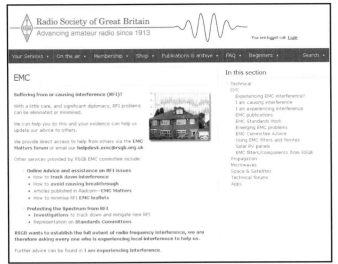

Fig 8.3: RSGB EMC Committee homepage www.rsgb.org/emc

Picture 8.3: Ferrite rings come in a variety of different sizes

electronic devices this problem will grow and needs to be controlled. Fortunately we now have legislation (British Standards Institute immunity requirements) which states that any apparatus should function satisfactorily in its environment and not cause undue electromagnetic disturbance to other equipment. This is a two way process;

Our equipment must have a certain level of immunity to radiation as well as not, itself, being a source of radiation.

We can expect new electronic equipment to meet these standards but older and poorly installed equipment may not.

Earlier we learnt the **immunity** of a device can often be improved by screening and filtering all power, signal and control leads. We also need to be aware of the **Routes of RF Entry** into a device. These have largely been covered already but there are two important cases,

Direct pick-up is where the circuitry of the device acts as an antenna. This type of interference tends to be independent of the frequency of the radiation.

Masthead and down-lead TV amplifiers are broad band and by definition amplify a wide range of frequencies which will include amateur transmissions. This may result in overloading the amplifier and or TV input.

Modern digital apparatus is subject to interference; DAB radio may have a loss of signal or muting and digital television pictures may freeze, pixelate, become jerky or disappear. Some other common causes of interference and their effects are given in **Table 8.1**

Some common methods to tackle interference

- Filters in mains leads from the power supply to the transmitter will prevent RF entering the mains wiring
- Ferrite rings on antenna down-leads and mains leads will minimise RF entering equipment.
- High-pass and low-pass filters may reduce the level of HF and VHF amateur transmissions into other equipment
- Mains filters in the leads to TV, radio and audio systems may reduce interference from electric motors and thermostats.

Cause of Interference	Effect of Interference
From radio transmissions	
Speech, particularly from AM and SSB	Speech like sounds in analogue radio, audio systems and telephones
Speech, FM	Muted or reduced volume of wanted signal
From Domestic and commercial apparatus	
Arcing thermostats	Buzzing sounds on analogue receivers which can correlate with when the device is being used, in particular, the opening and closing of a thermostat.
Vehicle ignition systems	
Electric motors	
Computers and peripherals	
Switch mode power supplies	
Plasma TVs	
Very high bit rate subscriber line (VDSL equipment)	
LED lighting	
Solar Voltaic (PV) inverters	

Table 8.1: Common causes of interference and their effects

8: Good Radio Housekeeping

9: Harmonics and Spurious Emissions

Check your RF signal for unwanted emissions

It is a requirement of your licensing conditions that you check for unwanted emissions from time to time, looking for harmonics and spurious emissions from your transmitter. You should check that you are not causing any interference to your neighbours' electrical equipment or any other users of 'wireless telegraphy'.

A transmitter that has been well designed and built should not produce excessive harmonic radiation and having a low pass filter in your RF output should deal with any low-level harmonic or spurious signals that may be present. Your transmitter may be producing spurious signals which are unrelated to the transmission or maybe only present when the carrier is modulating. Therefore, you should check your transmissions regularly and carefully.

How to check

The easiest method to check is to use a receiver that covers the frequencies that interest you. If using a transceiver, you will need to use a separate receiver to listen for spurious transmissions while your transceiver under test is in the transmit state. You may also want to ask another amateur some distance away from you to give you some help:

- First, connect a dummy load to the transmitter and turn the microphone gain to zero or unplug the microphone if that is easy to do.
- The receiver should be placed a short distance away, set to receive Morse (CW) and tuned to the same frequency as the transmitter, but without any antenna connected.
- Switch the transmitter to CW and transmit whilst checking the receiver. A strong signal should be received.
- Connect a well-screened dummy load to the receiver's antenna connector (a low power one will do fine) and check to see that the received signal level remains less than full scale on the receiver's signal strength meter. Alternatively, move the receiver further away and then see if the signal drops. This is one of the tests where you might want to check with another local amateur.

This is important to check, as if the received signal is too strong, it will overload the receiver. The overload in your receiver's circuits can produce harmonics and if this happens you will not be able to tell if the harmonics that you can detect are coming from the overloaded receiver or a faulty transmitter.

If the receiver shows less than an S9 signal on your meter, then it is unlikely to be producing its own harmonics, any harmonics or spurious signals you detect can be 'blamed' on the transmitter.

You may, of course, have access to more advanced RF test equipment, so feel free to substitute that here. The ideal piece of test equipment for looking for spurious emissions is a spectrum analyser.

Checking for harmonics

You will recall that harmonics can cause interference to other amateur bands, for example the second harmonic of a signal on 7.050MHz will be on 14.100MHz where amateur beacons transmit. However, not all bands have harmonics that fall into other amateur bands: The second harmonic of 50.50MHz will be on 101MHz, inside the VHF FM broadcast radio band.

You can check for harmonics as follows:

- Calculate the frequencies of the harmonics, that is 2, 3, 4 and 5 times the transmitted frequency, and make a note of them. Figure 9.1 gives an example.
- Tune the receiver to each of those frequencies and listen for any signals from your transmissions. It is quite possible that you will hear something, but it should be very weak.

If it is strong, move the receiver further away until the received level of the transmitted frequency is below full scale on the receiver's signal strength meter and try again. If the harmonics are still almost as strong as the fundamental it suggests a possible problem with the transmitter, in which case it is time to get help from a more experienced amateur or speak to your equipment supplier.

Interference from modulation

Within your Foundation studies you were advised on the importance of setting your microphone gain, Terminal Node Controller (TNC) audio output, or PC audio output level to keep your signal free from distortion so as not to cause interference to other radio users.

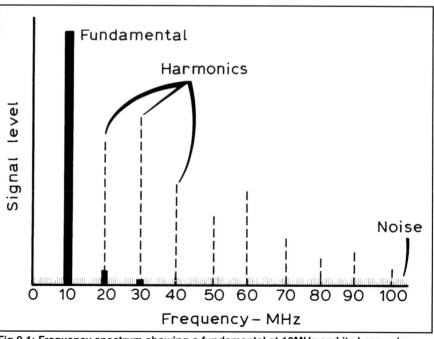

Fig 9.1: Frequency spectrum showing a fundamental at 10MHz and its harmonics

Overmodulation is caused by excessive audio amplitude in your AM transmitter; in turn this will cause your modulated signal to become distorted and make the bandwidth of the signal wider than the usual 6kHz. This causes interference to other stations on adjacent frequencies and will make you unpopular on air! This also applies to SSB.

Overdeviation is caused in an FM transmitter when the amplitude of your audio frequency causes a waveform to deviate more than is acceptable. This means that your transmitted signal becomes unintelligible and results in interference to other radio users. Apart from being unsociable on air, it is also a breach of your licence conditions.

Setting the microphone gain control or PC sound output too high (too loud) on an AM, SSB or FM transmitter could cause the microphone amplifier within the transmitter to produce audio frequency harmonics making the bandwidth to the audio feed to the modulator much wider than it should be. A general rule when operating digital modes is to reduce the audio drive to the transmitter to the lowest possible level to produce the desired output power. This will correspond to the lowest Automatic Level Control (ALC) meter reading, if your transmitter can show ALC. With speech via the microphone, the microphone gain should be adjusted such that the ALC meter shows speech peaks within the normal ALC level. These modulation methods are explained in more detail at the Full Licence level, but at this level it is sufficient to recall that applying excessive audio amplitude or excessive audio bandwidth to a modulator can cause excessive AM/SSB bandwidth or excessive FM deviation and that both can cause interference to adjacent radio frequencies

Checking for overmodulation

If it is possible, do some tests with other modes such as SSB and AM. If all is fine with these tests, odds on that FM will also be fine. For overmodulation and overdeviation tests, you will need to correctly readjust your microphone gain.

Whilst on SSB, quite a critical test is to try off tuning the receiver by up to 12kHz either side of your transmission. You must *really* be sure that your receiver is not being overloaded for this test. It must be comfortably off full scale on the received signal strength indicator. It would be most helpful to have the assistance of another local radio amateur to assist with this test.

Ideally, once you are more that 3 or 4kHz away from the transmitted frequency nothing should be heard. If, however, you can hear weak voice like signals you could try reducing the transmit power slightly and also the microphone gain. If one of these adjustments causes the received signals to come and go suddenly, you have reached the point where the transmitter is overloading or the modulator over-driving. If you do not feel confident with this, ask a more experienced amateur to help you.

If you have access to more extensive radio test equipment available, you may (of course) substitute it where appropriate.

Interference from CW

Not all transmitter interference comes from voice transmissions. Problems can also arise from poor CW keying. As you will recall, Morse code is sent by switching a continuous RF signal on and off to form the character's various dots and dashes. If the rise time of the RF power when the key is pressed down (or the fall time of RF power when the key is let up) is too rapid, a 'square' shaped RF envelope will be produced which is prone to causing interference.

Such a square waveform is likely to generate a signal using excessive bandwidth, with harmonics of the keying waveform extending across the band. The audible interference caused to other receivers sounds like clicking at the keying frequency and is known as key clicks.

In a good CW transmitter, a filter will be built into the keying stage. This filter will slow the rise and fall times of the RF envelope to maintain the narrow bandwidth of a good CW transmission.

Checking for spurious emissions

The spurious emissions from a transmitter can be at any frequency, and it can be quite difficult to find them. The best way to check this is to use a receiver to tune around the bands. However, strong broadcasting signals could be picked up despite not having an aerial connected. The best way to test would be to ask another radio amateur to assist by pressing the key as if sending Morse. By doing it this way it is more obvious if the signal is related to your transmitter as the received signal will come and go in time with the keying of the transmitter.

Modern radios offering a real-time wa-

terfall-style spectrum display can be great for this job, since you can see what on the display changes in time with the transmitter key position. The ideal tool for this job is a spectrum analyser, since this can display huge continuous spans of radio spectrum – while not every amateur will own such a device, many do, and they may be willing to assist you in making your measurements. If you are a member of a radio club, they may well have test and repair facilities and would be able to test your equipment much faster. They may have a spectrum analyser with someone comfortable using it. Such clubs often put on periodic events to test your equipment, and so these are a great opportunity to test your station.

Another option could be to use a cheap RTL-SDR USB dongle and suitable computer software which can show large continuous 'chunks' of radio spectrum – effectively a poor-man's spectrum analyser.

After the Tests

It is advisable to make notes during and after any tests, as they will be of good use if you ever need to ask a more skilled radio amateur or Ofcom for assistance later. Each time you conduct tests on your station, it is a good idea to archive such test results for reference at a later date. Keeping a record of your tests would be considered 'good practice' and keep you in good stead if you ever have a visit from an official Ofcom representative. You can also compare results over extended periods of time. Once you have made your measurements a couple of times, you will begin to find them easier and less time consuming.

10: Antenna Matching

The antenna and its feeder are vital parts of an amateur radio station. The best transmitting and receiving equipment in the world would be almost useless if it were not connected to a good antenna and the best antenna in the world would be equally useless if the RF signals are not carried to and from your radio along an efficient feeder.

An effective antenna system is something that you will have to create for yourself and what is best for you will depend to some degree on your own circumstances. Everyone's garden is different, and your constraints are not the same as anyone else's.

Even if the rest of your station consists of commercial equipment, your antenna is more likely to be home made. It is therefore important that you know some of the theory, so that you can put up the most effective antenna for your location. It will also be useful for the exam!

A little revision

Doubtless you will remember from your Foundation training that one of the most commonly used antennas is the half-wave dipole (see **Figure 10.1**). This is more often referred to simply as 'a dipole'.

Radio amateurs probably use the dipole more than any other antenna on the HF bands. It is also the basis for many VHF and UHF antennas, such as the Yagi. After a few simple calculations, a dipole can be constructed easily. It is simple to 'feed' and you can be working stations on the air in a matter of an hour or so – even sooner once you become proficient at making them!

Recalling your Foundation assessment, a dipole must be adjusted; trimmed to the correct length. Because we are talking about a 'half wave' dipole, you would be correct in guessing that this antenna should be one half of the wavelength in use. Factors such as if the wire is coated with insulation, or the closeness of the ground affect the exact length required in reality, so it is common practice for amateurs to make the antenna too long to begin with, and cut equal lengths off each end until a good SWR reading is found.

A transmitter needs a load

Transmitters are designed to transfer energy into a load. The antenna system (i.e., the antenna and its feeder) normally provide that load. If there is no load, or a load that is not what the transmitter is designed for, the power amplifier in the transmitter can be severely damaged.

Most modern amateur transmitters are designed to transfer energy into a load with an impedance of 50Ω. If the impedance varies too much from this value, the transmitter could be damaged. Some transmitters have an automatic power-reducing circuit to avoid damage caused by incorrect matching, but you should not rely on this. Consider this feature more an emergency protective measure than something to be used regularly.

A 50Ω resistor can be used as a load for a transmitter to work into, for example, as a dummy load for test purposes. In this case, the resistor must not be of the 'wire wound' variety, since this coiled wire construction would have the effect of adding inductance. The resistor used must also be capable of handling the transmitter output power without failing.

In an antenna system, instead of the transmitter power being converted into heat as in the dummy load, the power is transferred through the feeder to the antenna to produce a radiating radio wave.

At this level, we are not required to investigate the theory in any detail – it is quite complex as you will imagine. It is simply enough to know that a properly tuned antenna system will only provide the correct load if we feed it from a transmitter at the correct frequency.

Feeder Impedance

The first component in the antenna system is the feeder. The feeder is the cable that carries the RF signals between your radio and the antenna.

Several types of feeder are used by radio amateurs and each type has its own 'characteristic impedance'. This impedance is not the DC resistance of the wire but an AC characteristic; where there is resistance and reactance present in the same circuit you have impedance. In a feeder, the wire has some resistance and the two conductors form a kind of elongated capacitor that introduces some reactance. The lengths of wire themselves also have an associated inductance; these factors combined together give us an impedance. The characteristic impedance of coax is determined by the diameter and spacing of the conductors. Coaxial cables for amateur radio are normally 50Ω impedance, to suit modern transmitters. The coax used for television down-leads is normally 75Ω, but there are several other impedances available. If you buy a length of feeder, you should make sure it has the characteristic impedance that you want.

If the coax is correctly terminated (that is, terminated by a resistive load that is equal in value to the characteristic impedance of the coax) the length of the coax will have no effect on the impedance – if there is a 50Ω load at one end, the impedance (Z) will be 50Ω at the other end.

You should also note that the characteristic impedance determines the ratio of the RF RMS potential difference to the RF RMS current in a correctly terminated feeder. Any calculations required on this topic would be the same as the other impedance calculations covered earlier in the book:

$$Z = \frac{V_{RMS}}{I_{RMS}}$$

The next part of the system is the antenna itself, and we see that antennas themselves also have impedance.

Shown in **pictures 10.1** and **10.2** is a typical dummy load used by amateurs. It shows the maximum power loading as well

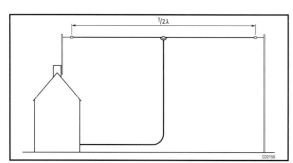

Fig 10.1: A dipole mounted horizontally in the garden

Picture 10.1: A typical dummy load showing the maximum power loading

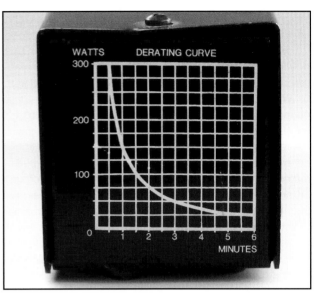

Picture 10.2: A dummy load graph indicating how long a known power output may be applied

as a graph indicating how long a known power output may be applied.

Feedpoint Impedance

Antennas are generally formed from metal wires, rods or tubes, all of which have some resistance. The antenna elements are conductors separated by air, forming a kind of capacitor, which has some reactance. Some antennas include coils, which also have some reactance. All antennas therefore have impedance associated with them. All these separate impedances combine into a feed point impedance for each antenna.

You should recall that a half-wave dipole has two sides or elements. When it is cut to the correct length for the frequency in use, the impedance at the centre of the dipole in free space is 73Ω. Due to the presence of nearby objects, and the effect of proximity to the ground, it is somewhere closer to 50Ω in a practical setting. Different antennas have different feed impedances.

The feed impedance of an antenna is not fixed; it is related to the physical dimensions of the antenna and the wavelength of the signal you are feeding it with. As the length of an antenna is changed to suit the wavelength of the intended signal the feed point impedance changes. When the length of the antenna is correct the feed point impedance is close to the 50Ω that the transmitter was designed to use.

Taking this a little further, you should be able to see that if you apply an RF potential difference across the antenna feed point impedance, some RF current will flow in the antenna. Following what we have already learned about Ohm's law in AC circuits it should be no surprise to find that the RF current (in Amps) flowing in the antenna is related to the feed point impedance (in Ohms) and the potential difference of the applied signal (in Volts). In general, you will get maximum radiation from the antenna when you have maximum RF current flowing in the antenna elements.

Balanced and unbalanced Antennas

You will recall that in the Foundation Course twin feeder was described as being **balanced** and coax feeder as being **unbalanced**. We need a better understanding of what happens at the feedpoint of an antenna. The centre core of the coax carries a current which is conducted along one arm of the dipole. There is nowhere else for the current to be conducted. The braid of the coax is connected to the other arm of the dipole. We would expect all the current to be conducted into this arm of the dipole but as can be seen in **Figure 10.2** because of a peculiarity in which RF currents can be conducted some of it can travel back down the outside of the coax braid. In reality, conduction to the antenna is on the inside wall of the braid and some current returns on the outside of the braid. This presents us with two problems. Firstly, the current in both arms of the antenna is not equal and secondly there is RF being radiated from the feeder. You will be aware "that RF in the shack" is very undesirable.

The ideal situation is to have equal and opposite currents flowing in the centre core and inner wall only of the braid giving us **differential mode currents**. The reality is that we have current on the outer braid flowing in the same direction as that in the core giving us **common mode** currents.

The problem can be resolved with a current balun or choke balun such as described by winding feeder cable round a ferrite ring. This is nothing more than an

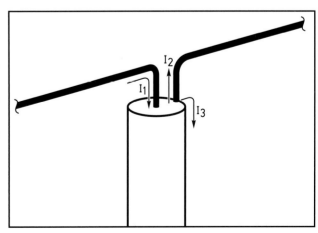

Fig 10.2: RF travel in a feeder
I_1 - Current flowing in the inner core
I_2 - Current flowing on the inside of the braid
I_3 - Current flowing on the outside of the braid

10: Antenna Matching

inductance to prevent or "choke off" the common mode currents.

Twin feeder, usually called ladder line or window line can also be affected by common mode currents.

When twin feeder is connected to a perfectly balanced dipole we expect the currents flowing in the two wires taking the received signal to the transceiver to be exactly equal and opposite. We must remember that the feeder itself acts like an antenna and radio signals passing through it will induce a current in the same direction in both wires. We call these unwanted currents common mode currents. Its effect is shown in **Figure 10.3**.

The common mode currents can be suppressed by passing the twin feed through a ferrite ring. The inductive effect of the ferrite ring suppresses the common mode currents.

The ferrite ring needs to be fitted around the twin feed as near to the transceiver as is possible.

Antennas come in different shapes and sizes and will therefore present different impedances to the feeder line. Baluns can be designed to act as impedance matching devices. This idea will be developed in the Full Licence

Matching

We now know that modern transmitters are designed to transfer energy into a load of 50Ω, and that commonly-used coaxial cable has a characteristic impedance of 50Ω, and we have just discovered that a dipole has a feed point impedance of about 50Ω when it is used at its designed frequency. If the transmitter, feeder and antenna all have the same impedance, we say that they are matched. This is the ideal case.

The benefit of using a matched system is that no matter how long the feeder is, the impedance at the transmitter will always be the same as the impedance at the antenna, just as if the feeder were not there. This is also when the feeder has the lowest loss. In such circumstances the transmitter will not be damaged, and we will have the most efficient transfer of energy from our transmitter to the antenna.

A dummy load can be a useful tool in fault finding, since they present a good SWR to the transmitter. The dummy load can be connected to the remote (antenna) end of a feeder and the SWR observed at the transmitter side. If the dummy load presents a poor match to the transmitter, it is possible to conclude that the feeder is affecting the dummy load impedance, and that the cable may be faulty.

But what happens if there is a mismatch? There are two points where this can happen, between the transmitter and the feeder and at the junction of the feeder and antenna.

Mismatch at the antenna

If an antenna is not adjusted to the correct length for the frequency in use, some of the RF power arriving through the feeder from the transmitter will be reflected back down the feeder, as **Figure 10.5** shows. The reflected power is not lost, but it will combine with the power travelling up the cable to form what are called standing waves. These are points of high and low voltage, and if you could see the pattern they form, it would look a little like stationary waves along the cable. You will have measured the standing wave ratio (SWR) as part of your Foundation assessment.

The reason that some of the power is reflected is that if the antenna is not the right length for the frequency in use, the feed point impedance will no longer be 50Ω. The antenna will only present the correct impedance when the antenna's length is correct for the wavelength of the signal being used. Maximum transfer of energy can only occur when the impedance of the antenna matches that of the feeder, so if there is any mismatch some power will be reflected.

The proportion of power that is reflected depends on how much of a mismatch there is. When using a feeder of 50Ω impedance, we might have an antenna with an impedance of 70Ω. This is reasonable and would actually be considered an acceptable match. On the other hand, if the feed point impedance of an antenna were 500Ω, this would be a bad mismatch.

The amount of standing waves depends on the size of mismatch. The greater the mismatch, the greater the standing waves are. When we measure the SWR on a feeder we are, in effect, find-

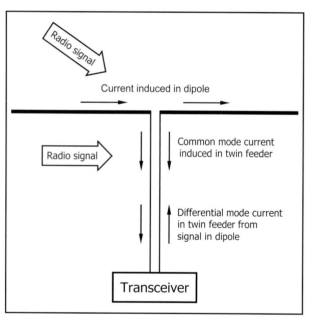

Fig 10.3: Differential and common mode currents in a twin feeder

Fig 10.4 Currents flowing in the twin feed are represented by arrows

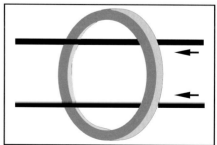

A: Common mode currents

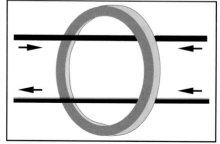

B: Differential mode currents and common mode currents

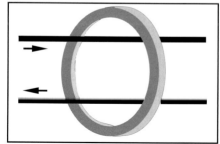

C: Differential mode currents following suppression of the common mode currents by the ferrite ring

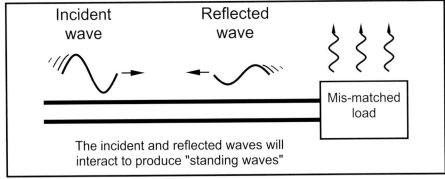

Fig 10.5: Standing waves on the feeder of an antenna

ing out to what extent our antenna system is mismatched. A lower SWR means a better match as there are fewer standing waves present on the feeder.

It is important to note that the reflected wave is also subject to feeder loss, and as such poor (lossy) feeder can have the effect of hiding a poor antenna SWR since less reflected power is observed at the transmitter.

Mismatch at the transmitter

The main problem with having a mismatched antenna is that the reflected signal will change the input impedance of the feeder so that it is no longer the characteristic impedance and the feeder will not then present the correct load to the transmitter. Instead of the 50Ω of a well-matched system, the transmitter will instead be connected to some other impedance. You don't need to know how to calculate this and it is enough to say that the transmitter will not operate at full effectiveness if there is a mismatch between the transmitter and the feeder. In extreme cases, this can permanently damage the transmitter, although this is uncommon with modern radios.

Most transmitters can tolerate a small amount of mismatch, and as such, we can successfully use a single antenna over a whole amateur band. When we change frequency within the band the match between the antenna and feeder will change because the wavelength will be slightly longer or shorter, but the antenna stays the same length. As a result, the level of standing waves will change, and the transmitter will 'see' different impedances at its antenna socket. However, because the mismatch will be quite small, the transmitter should be able to cope without difficulty.

Antenna Matching Units

On the HF bands it is common practice to use a single antenna on several different bands. Such an antenna will only be perfectly matched at one spot frequency and will normally only present an acceptable impedance across one band. On all the other bands the rather large mismatch will result in a high SWR. If this is the case, you can use an antenna matching unit (AMU) between the transmitter and feeder.

An AMU contains variable inductors and capacitors. These variable inductors and capacitors have associated impedances, as we have seen. These impedances inside the AMU can be combined with the antenna impedance at the transmitter end of the feeder to make the mismatched antenna impedance more acceptable to the transmitter. This will allow the transmitter to feed power through the feeder into the antenna. Many amateurs will refer to this apparatus as an Antenna Tuning Unit (ATU); it is more correctly called an Antenna Matching Unit. The position in which an AMU would be used in your station is shown in **Figure 10.6**.

An AMU connected between the transmitter and the feeder will not remove the mismatch between the antenna and the feeder. Neither will it change the level of standing waves on the feeder. What it will do is let the transmitter 'see' the correct impedance and so feed the maximum power into the antenna system.

You should note that an AMU connected to a transmitter as shown in Figure 10.6 does not change the feed point impedance of the antenna, so although you may have a 'perfect' SWR reading at the transmitter, the antenna may not radiate very well. Remember that a dummy load presents an excellent match, but it does not radiate RF energy very well! Even when working well, the AMU may introduce a significant loss, so an antenna system with an AMU will never be as good as a correctly cut and tuned dipole. Do not be tempted to trade the ease of using an AMU over correct antenna adjustment.

One final remark to be aware of: when the transmitter output is correctly matched with an AMU, the AMU may provide additional protection against the radiation of harmonics. However, this shouldn't be relied on to suppress unwanted spurious signals.

Further Information

If you would like to learn more about antennas, there are many sources to refer to. The *Backyard Antennas* book by the RSGB comes highly recommended, containing designs for building antennas, together with some of the technical background.

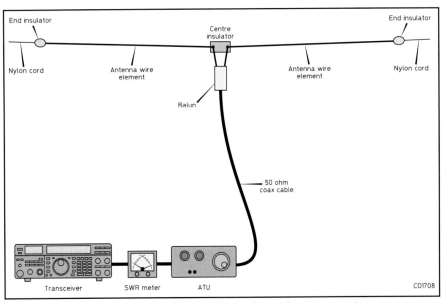

Fig 10.6: Layout of a transceiver, SWR, ATU, feeder, and antenna system

11: Feeders and Baluns

Feeders

Feeders and Currents

A Feeder is the common name for what is more correctly known as a Transmission Line. This is a cable of special design that is able to conduct alternating current of Radio Frequency. These frequencies are high enough that their wave nature needs to be taken into account.

On transmission radio frequency currents pass along the feeder and energy radiates from the antenna as electromagnetic waves. Conversely, when electromagnetic waves interact with the antenna a Radio Frequency current is induced which passes along the feeder to the transceiver.

The electromagnetic wave and the feeder can be considered to be of comparable length and this leads to an interesting phenomenon; the current and voltage along the feeder, at any point in time, will also vary in the same sinusoidal manner as the electromagnetic wave.

Figure 11.1 shows the variation in current (or voltage) along the feeder for one complete wavelength. Let us look at the mid-point, X, where the current is zero. It follows that at the same time the current will also be zero half a wavelength up the feeder and half a wavelength down the feeder. Similarly, a quarter of a wavelength up the feeder the current will show a positive peak and a quarter of a wavelength down the feeder there will be a negative peak.

If the wavelength of the received signal was 20m, at any point in time, we would have zero current every 10m along the feeder. Also, there would be alternating maxima and minima currents at 10m intervals.

Coaxial Cable

During your Foundation studies you will have learned about coaxial cable, commonly called coax. **Picture 11.1** shows the construction of coax. It contains an inner and outer conductor. The inner conducting core is generally thick copper wire and the outer conductor is a braid made from thinner wire. They are separated from each other by a sheath of insulating plastic, sometimes called the dielectric. Around the braided wire is another layer of insulating plastic on which the manufacturer will print the cable specifications. The braid is connected to ground in each piece of equipment.

Coax is said to be an unbalanced feeder. Provided the cable is correctly terminated, with good low resistance electrical connections between all the plugs and sockets the signal is unaffected by nearby objects. This is why coax is popular, since it can be easily routed round other equipment, laid on the ground or with other cabling, or even fixed to a metal mast without concern for the signal being carried.

Coaxial cable comes in a variety of types with varied thicknesses (see Picture 11.1). If you have bought coax, you may have been advised that you needed the 50Ω variety. This resistance does not refer to the wire, but to the **characteristic impedance** of the cable. The characteristic impedance must be matched to the impedance of the antenna and transmitter, such

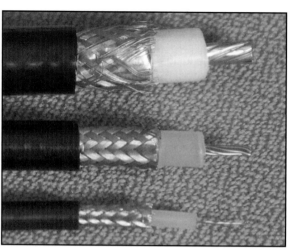

Picture 11.1: A good quality coaxial cable has a thick outside braid to contain the signal carried on the centre. Poor quality cable has a thinner braid with fewer wire strands

that all parts of the system have the same impedance, as discussed previously. 50Ω coaxial feeder is the most commonly used in amateur radio.

For television and satellite receivers coaxial cable of 75Ω impedance is used.

The characteristic impedance of coax cable is dependent on the diameter of – and spacing between – the conductors making the cable. The coax impedance is typically printed on the outer jacket of the cable to avoid confusion.

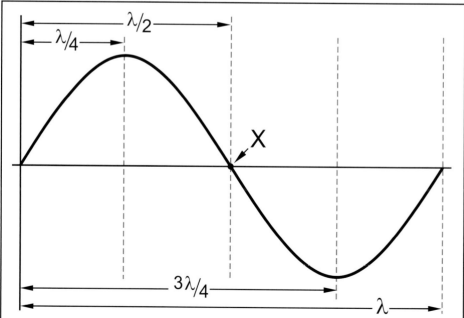

Fig 11.1 A graph showing the sinusoidal variation of current along a feeder at a specific point in time.

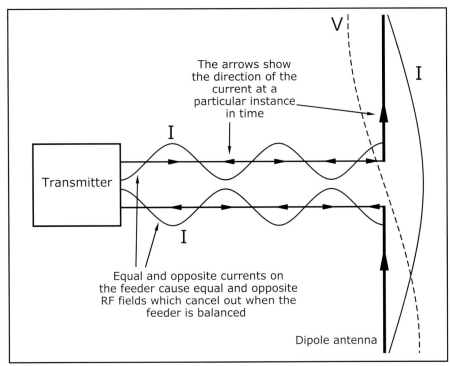

Picture 11.2: Balanced feeders are two separate wires held apart by plastic spacers at a fixed distance - they come in a variety of shapes and sizes, which governs their characteristic impedance

Fig 11.2: Equal and opposite currents in balanced feeder cancel out to stop radiation from the feeder

Balanced Feeder

Another common feeder type is balanced feeder, sometimes called twin feeder, twin lead, ladder line, or open-wire feeder. Those names alone hint at the construction and appearance of such cable! As you see from **Picture 11.2**, balanced line is made from two identical pieces of wire held apart at a fixed distance by insulators.

When used correctly, balanced feeder is fed with a signal such that the current in each of the two wires is the exact opposite – exactly balanced. It is reasonable to wonder why feeder does not radiate the signal, since balanced feeder clearly does not have the shield surrounding it to 'trap' the signal in like coax. To answer this question, recall that an alternating current flowing through a wire creates an electromagnetic field. RF currents are no exception, and this is the basis of how a long wire antenna radiates. However, since balanced line has two wires with exactly opposite currents, each of the conductors has an equal yet opposite field around it. Any radiating field from one conductor is exactly cancelled out by the opposite field from the other conductor and so the net radiation from the feeder is zero. **Figure 11.2** shows how the fields cancel in the feeder, allowing just the antenna to radiate.

This property of balanced line is only true if both currents balance. If the currents are not equal, then the fields will not balance, and the line will radiate – this is typically not desired. Unlike coax, nearby metal objects can also unbalance the electromagnetic fields causing them not to cancel entirely and cause the feeder to radiate. For this reason, balanced feeder is better suited to being in free space away from other objects.

As with coax, the characteristic impedance of balanced feeder depends on the diameters and spacing of the two conductors, and all of the same rules apply. Widely used balanced feeder impedances are 300Ω and 450Ω, but values between 75Ω and 600Ω are commonly available.

Waveguide

Coaxial cable and balanced feeder represent most of the feeder cables in use in an amateur radio station. However, there are several other types of feeder available which may better suit some scenarios. One such example is the waveguide. A waveguide is a hollow conductive metal pipe that is used to confine radio waves in a similar way to how a hose-pipe confines water. Typically, the cross section of the waveguide is rectangular, and the larger dimension must be greater than λ/2 for the signal to travel through. **Picture 11.3** shows an example of a piece of waveguide. Since the waveguide dimensions are set by the signal frequency, a waveguide becomes impractical to use at lower frequencies where the wavelength is large, and as such, a waveguide is most commonly found in microwave systems.

Since a waveguide is too large to be practical at lower frequencies, coax might seem to be the obvious choice for a feeder. However balanced feeder offers several advantages: it has lower loss than coax, can be connected directly to balanced antennas (such as dipoles) without the need for a balun, and can be less susceptible to EMC problems.

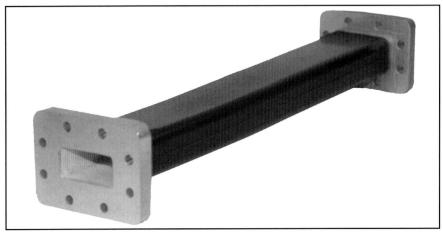

Picture 11.3: A section of waveguide showing the rectangular cross section where signals travel

11: Feeders and Baluns

Feeder Losses in dB

All feeders cause some loss in signal, but some are less lossy than others. Possibly the biggest advantage of using balanced feeder over coaxial cable is the reduced losses. As a signal travels through the feeder, some of the signal energy is lost, causing the cable to heat slightly. This is true on both transmit and receive. The longer the feeder the greater is the loss. The losses in balanced feeder are generally considerably less than those of coaxial cable. When used correctly, balanced feeders can run for hundreds of metres with virtually no loss at all. Different coax cables have very different losses, with losses typically increasing with frequency.

In science and engineering, losses (and gains) are most often given in decibels, which are abbreviated to "dB". The dB expresses how much smaller (or bigger) one quantity is compared to another, and feeder manufacturers will supply loss figures in dB. You do not need to know how to calculate dB at this stage, but you will be required to perform simple calculations with losses and gains using only multiples of 3dB and 10dB. **Table 11.1** shows some common decibel figures and the associated loss.

Key facts to remember about decibels are:
- A loss of 3dB corresponds to a halving of power
- A loss of 10dB corresponds to a tenth of the original power
- Decibels are added together in calculations, such that a loss of 6dB corresponds to a half of a half, or a quarter.
- A loss of 7dB can be expressed as 10dB – 3dB: this corresponds to twice the value of one tenth, or one fifth.

This is important because your licence conditions state the maximum power at the antenna. If you encounter significant feeder losses, you may not be making the most of your licence!

Examples

A manufacturer of coax tells you that the loss per metre is 1dB. Your feeder is 16 metres long. With 50W of transmit power, how much would you expect reaches the antenna? The 16 metres of coax would have 16dB of loss. Using just the key points above, we can break this down to 10 + 3 + 3 = 16dB. Taking simple steps, we can work out the power after each addition:
- After 10dB, we have only one tenth of our original power left, so 5W.
- After 3dB, we lose a further half, so 2.5W.
- After the final 3dB, we are left with 1.25W.

Clearly this would be a poor choice of cable.

A transmitter is set with an output power of 32W. At the antenna, 8W is measured. What is the feeder loss? To calculate this, work from 32W by halving the power (i.e. stepping down 3dB at a time) until you reach the result. Half of 32W is 16W (3dB), and half of 16W is 8W (3dB). The total feeder loss is therefore 3 + 3 = 6dB.

The versatility of the decibel

The decibel can be used in describing a change in power. In the example shown in **Table 11.2** there are no units, All we have done is expressed a change as a ratio. (More correctly, as a logarithmic form of the ratio) You will notice that very large awkward to manage numbers are simplified to a more convenient form for everyday use.

Decibels can be used to specify other electrical units

As an Intermediate Licensee you have the privilege of transmitting at 50 W output or 17dBW. In this case we are referring the power to 1Watt.

How does 50W get changed to 17dBW?

$$50 = 10 \times 10 \times 0.5$$
which becomes
$$10\text{dBW} + 10\text{dBW} - 3\text{dBW} = 17\text{dBW}$$

A Full Licensee can transmit at 400W or 26dBW. Again, we are referring the power to 1 Watt. How does 400W get changed to 26 dBW?

$$400 = 2 \times 2 \times 10 \times 10$$
which becomes
$$3\text{dBW} + 3\text{dBW} + 10\text{dBW} + 10\text{dBW} = 26\text{dBW}$$

Sometimes 1 Watt may be considered a very large value to use as a reference

Loss (dB)	Fraction of power remaining
3	One half
6	One quarter
7	One fifth
9	One eighth
10	One tenth

Table 11.1: Conversion between loss in dB and remaining signal power

and 1mW or 0.001W is preferred. We would see the value stated in dBm where the 'm' is to remind you that the reference level is 1mW. How would we express 100W in dBm?

$$100\text{W is } 10 \times 10 \times (10 \times 10 \times 10) \text{ mW}$$
which becomes
$$10\text{dBm} + 10\text{dBm} + 10\text{dBm} + 10\text{dBm} + 10\text{dBm} = 50\text{dBm}$$

(To avoid confusion in the calculation, (10 x 10 x10) reminds us there are 1000mW in 1W)

Later, you will also see the importance of using dBi when comparing the power output of the dipole antenna to that of the theoretical isotropic antenna.

Calculations become simpler when using decibels

This example is beyond that required for the exam but it will help you to understand the versatility of the unit and give you a little more insight to its use in the hobby. Suppose we have the following information and we wanted to know how much power entered another amateur's receiver.

Power output of transmitter	50W
Cable loss in feed to transmitter antenna	-12dB
Gain of transmitter Yagi antenna	+16dB
Loss of power in radiation from transmitter to receiver dipole	-62dB
Gain of receiver dipole	+4dB
Loss in power in receiver feeder	-6dB
total loss in power = -12 + 16 - 62 + 4 - 6 or -60dB	

In this example -60dB tells us that the power has been reduced to one millionth, or 1×10^{-6} of the power leaving the transmitter. The power arriving at the receiver must be $50 \times 1 \times 10^{-6}$W or 5×10^{-5}W

Decibels and the S meter

Decibels play an important part in the understanding of an S meter scale. The S-meter is also scaled in dBs. One S-point, from S1 to S2, or S7 to S8, for example is a doubling of the voltage. Remembering that power is proportional to V^2 there is a 2^2 or fourfold increase in power which is a 6dB increase.

RSGB Book Shop

Always supplying the best Antenna books

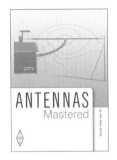

Radio Society of Great Britain
www.rsgbshop.org
3 Abbey Court, Priory Business Park, Bedford, MK44 3WH
Tel: 01234 832 700   Fax: 01234 831 496

FREE P&P on orders over £30. See T&Cs

Picture 11.4: A common design for a choke balun is to wind coax around a former

The Balun

We have already discussed balanced feeders such as 'ladder line' and unbalanced feeders such as coaxial cable. A balun can convert from balanced-to-unbalanced feeder, and vice versa. Baluns are often found at the interface between balanced feeders and coax. Another common use of a balun is at the centre of a dipole. A 4:1 or a 9:1 balun may also perform the task of impedance matching. As a practising amateur you may decide to try an antenna called a Delta Loop. Delta loops are not part of the Intermediate syllabus but they are an example of an antenna not having a 50Ω impedance and where a balun will be necessary.

Since the dipole is a balanced antenna it requires a balun to convert the balanced currents in the radiating dipole wires to unbalanced for connection with coax. Without the balun, RF current will try to conduct along the outer braid of the coax, causing EMC issues since the signal is no longer confined. This also makes the dipole harder to match. The coax may also conduct noise along the outer braid of the coax into the antenna. The specific type of balun required in this scenario is called a choke balun.

Choke baluns can be constructed in various ways. A common method, often called the 'ugly balun' simply coils the coax feeder around a former (a plastic soft drinks bottle or piece of plastic water pipe) such as in **Picture 11.4**. From the Basic Electronics chapter, you will recall that this creates an inductor and that the inductor opposes the flow of AC currents. Another common balun construction technique is to put several loops of coax through a ferrite ring or clip-on ferrite shell. These ferrite cores considerably increase the inductance of the choke, improving their effectiveness on lower frequencies. You are reminded of the use of ferrite rings to remove common mode currents in Chapter 10.

Example	Change expressed in words and as a ratio	Change in dB
Power Amplifier 10W to 20W	Twice or x 2	+3dB
Radiated power at a Yagi antenna 10W to 100W	Ten times or x 10	+10dB
Feeder loss 100W to 25W	A quarter; a half of a half, or 0.5 x 0.5 = 0.25	-6dB
Reduction of signal in a filter 1W to 1 x 10^{-6}W	One millionth or x 0.000,001	-60dB

Table 11.2: Examples of the use of decibels

12: Antenna Concepts

Antenna Gain in Decibels

You will recall from your Foundation studies that antennas have a property known as gain. This gain is a measure of how effectively the antenna can concentrate or focus the RF energy into a given direction. To allow us to compare antennas and judge one's performance against another's, a reference antenna is useful.

An isotropic antenna is a theoretical type of antenna that radiates RF energy with the same intensity in every direction – it cannot physically be made – but it is a good example to compare other antennas to. If you were to transmit into this theoretical isotropic antenna, you would measure your power with **Effective Isotropic Radiated Power (EIRP)** in dBi. Notice how, since the dB is now relative to an isotropic antenna, an 'i' has been added to the end. It is important to say that the EIRP gain is 0dBi. This means if it was possible to transmit 10 Watts into an isotropic antenna, you would radiate a signal with 10 Watts EIRP.

Since the isotropic antenna only exists in theory, another common reference antenna is the half-wave dipole. A dipole does not radiate RF energy from the ends of the wire, and the radiated energy is slightly stronger in the antenna's optimal direction. The half-wave dipole has a gain of 2.15dBi. You are expected to recall this. it is also useful to note that antenna gains are often given relative to a dipole instead – these gains typically state units of dBd, with a 'd' at the end, to indicate a dipole reference. As before, the conversion factor from dBi to dBd is to add 2.15dB.

$$dBi = dBd + 2.15$$

When we consider gain, 3dB represents a doubling of power and 10dB represents a factor of ten times, just as when we looked at losses in coax. If we consider a station transmitting 10W into an antenna with a gain of 6dBd, then their effective radiated power (ERP) is 40W in the optimal antenna direction. If we were to have an antenna with a gain of 10dBd and set our transmitter to 50W, then the ERP would be 500W when compared to a half-wave dipole! This is perfectly valid to do, since the licence schedule typically states the maximum power to the antenna – but be careful to check the schedule, since on some bands there are also radiated power limits!

The final thing to note with antenna gain is that the benefit is also present on receive. An antenna with higher gain should give you a stronger receive signal and is less receptive to signals in other directions.

Polar Diagrams

When we discussed antenna gain, we used an isotropic antenna which radiated equally well in every direction, every side, top and bottom, like a perfect sphere. However, as noted, this is a theoretical idea only, and cannot exist in reality. Most antennas do not radiate equally well in all directions. A vertical antenna comes close, since it radiates energy in all directions around the vertical part, but it does not radiate directly above or below.

You may have seen TV antennas on rooftops all pointing towards the television transmitter. These antennas have a large gain in a certain direction, and so must be correctly aligned. The half-wave dipole is a common antenna used in amateur radio, and it too, has gain compared to an isotropic antenna. The dipole is generally run lengthways down a garden, suspended by one or two supports. When used in this way, the dipole is directional, with the maximum gain being at right angles to the wire elements – if your dipole wires run north to south, the maximum gain will be east-west.

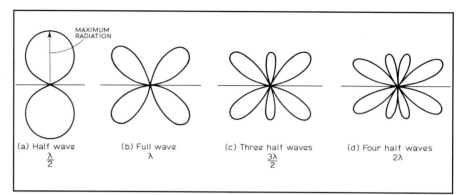

Fig 12.1: The polar diagrams of dipole antennas of varied lengths. You are only expected to recognise the half-wave dipole.

Figure 12.1 shows the polar diagrams for dipole antennas of various lengths. Looking at the half-wave dipole, (a), it is clear to see that there is fairly broad pattern of radiation (called 'lobes') to each side of the wire, which reduce as you move towards the ends of the wire. Note that these diagrams show only the two-dimensional view, but that this pattern is wrapped around the wire in three dimensions. In real life nearby objects will disrupt the field around the antenna, so the pattern will vary from the ideal.

The Yagi Antenna

To better understand how the Yagi antenna works, we can consider a few simple steps to help give you an idea of how they are constructed.

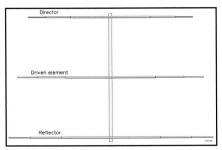

Fig 12.2: A three element Yagi antenna

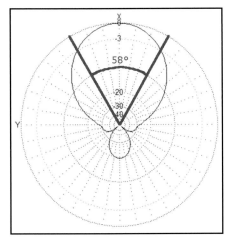

Fig 12.3: The antenna beamwidth shows how tightly focused the antenna beam is

First consider a simple half-wave dipole with the radiation pattern as shown in Figure 12.1 (a).

You can see that the dipole radiates energy from either side of the wire, i.e., up and down the length of this page. Now imagine we could somehow mirror the energy coming down the page back up towards the top of the page so that twice as much energy went in one direction and none went in the other. This is the job of the reflector, an element slightly longer than the driven (dipole) element that mirrors the RF back towards the driven element and towards the front of the antenna. In practice, this mirroring is not perfect and some RF energy still leaves the antenna from behind. One measure of an antenna's performance is the ratio of energy leaving from the front of the antenna to that leaving the back, this is known as the **front to back ratio** and is usually expressed in dB.

At the present, we have a 2-element antenna: a driven element (our ½ wave dipole) at the front and the slightly larger reflector at the back. Adding directors to the front of the antenna has the effect of focusing the RF energy into a smaller area – they direct the transmitter energy. A Yagi has one or more directors, which are shorter than the driven element. The antenna's beamwidth is usually given as the angle between the half-power (-3dB) points of the main radiation lobe. This is shown on the antenna's polar diagram in **Figure 12.3**. Here, the beamwidth is 58°.

As the number of director elements increases, the antenna better focuses the RF energy into a narrower beamwidth, and so it is said that the antenna's gain is higher in that direction. The gain of an antenna comes from concentrating the RF energy into a narrower beamwidth.

So far, we have only considered what the RF energy would look like if we could look at the RF from high above the antenna. An antenna is a 3-dimensional object and so you may also be interested in the angle the RF leaves the antenna, called **the angle of radiation**. The angle of radiation is measured between the ground (the horizontal) and the peak of the main radiation lobe. **Figure 12.4** looks at the Yagi from the side and shows how the RF leaves the antenna – here the angle of radiation is 23°.

A low angle of radiation is needed for working long distances, and the angle of radiation is affected by the antenna's height from the ground – the closer the antenna to the ground, the larger the angle of radiation. This is true for all antennas, not just the Yagi, and it is why antennas work better when they are elevated well above the ground in free space.

Designing and understanding the characteristics of an antenna has become a lot easier with the personal computer and modelling software. **Figure 12.5** is taken from an antenna simulation performed by MMANA-GAL. The two images side-by-side show the antenna front to back ratio and beamwidth on the left with the angle of elevation on the right. The key on the right shows the key measurements for our simple Yagi.

Antenna modelling software can go one step further and show what this 3D radiation pattern looks like, combining both the beamwidth and the angle of elevation plots into one diagram. **Figure 12.6**

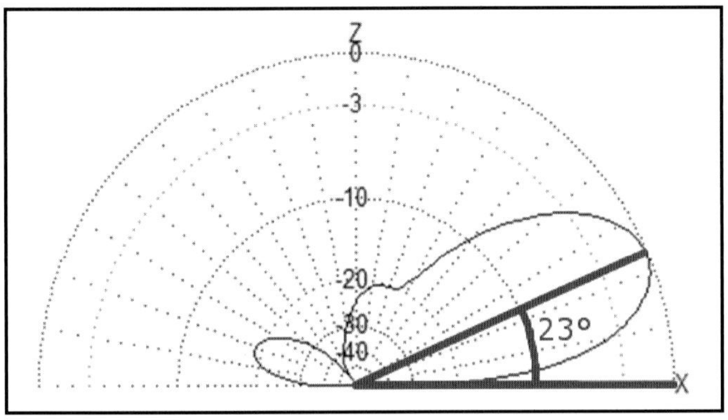

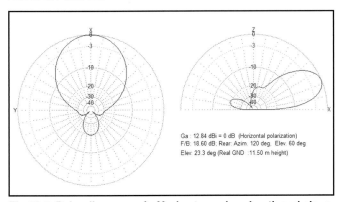

Fig 12.4: The angle of radiation shows what angle the RF energy leaves the antenna

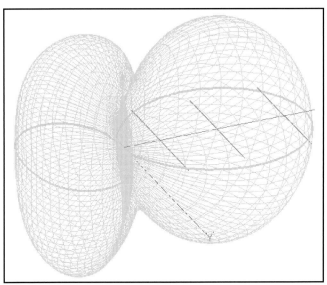

Fig 12.5: Polar diagrams of a Yagi antenna in azimuth and elevation (MMANA-GAL software)

Fig 12.6: 3D far field radiation pattern for a 3-element Yagi

12: Antenna Concepts

shows such an image, again taken from MMANA-GAL.

It is also worth mentioning (again) that this explanation is valid for receiving, too! Any improvement that an antenna makes on transmit is also there when receiving.

The Trapped Dipole

You should have a good understanding of how a half-wave dipole is constructed from your Foundation course. You will recall that the total length of the dipole is a half of the wavelength of the frequency for which it is tuned, so that for example, a half-wave dipole for the 40 metre (7MHz) band will be approximately 20 metres in total length. If we wished to instead use this antenna on the 20 metre (14MHz) band, the total length would need to be 10 metres. Each side would go from being 10 metres to 5 metres long.

The principle of a trapped dipole is to use frequency selective components to disconnect the extra antenna length when it is not required. In reality, the lengths will need to be adjusted slightly depending on the wire and traps used, but **Figure 12.7** shows the basic principle.

As can be seen in Figure 12.7, each trap is a simple parallel LC tuned circuit. To frequencies lower than the trap frequency, the RF current can travel through the trap to the wire element continuing on the other side. However, as the frequency increases, the RF current can no-longer go through the trap and so the RF current is 'trapped' to the inside (shorter) wires. Using this principle, a multiband antenna can be made that presents a good feed point impedance of approximately 50 Ω on more than one band.

It is also common to see traps used in other multi-element antennas, such as Yagis. Several manufacturers make HF multiband beams for the 20, 15 and 10 metre bands using traps in the beam elements. **Picture 12.1** shows one such example, where the traps are visible on the beam elements.

Polarisation

From your Foundation studies you should recall that the polarisation of a radio wave is governed by the orientation of the antenna, with a vertically mounted antenna producing vertical polarisation and a horizontally mounted antenna producing horizontally polarised waves. On HF, the effects of polarisation are rarely noticed because the ionosphere is constantly changing and the reflected polarisation is constantly twisting. However on VHF and UHF, it is important to match your antenna's polarisation to the antenna polarisation of the distant station otherwise a significant loss in signal strength is experienced as a result of cross-polarisation.

Radio waves are one type of electromagnetic radiation. Each type of electromagnetic radiation involves two fields which are at right angles to each other; an electrical field (denoted by the letter 'E') and a magnetic field (denoted by the letter 'H') - hence the term 'electro-magnetic' radiation. It is the electric field of a wave that defines its polarisation.

As RF current flows through a conductor such as a Yagi or wire dipole element, a magnetic field is produced proportional to the current, as we have seen when considering inductors and transformers. All magnetic fields have an associated electric field. The two fields are at right-angles to each other and are inseparable. The electric field is produced in the same plane as the antenna elements, and the magnetic field is produced at a right-angle to the antenna's elements. This can be seen in **Figure 12.8**.

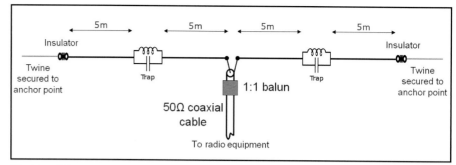

Fig 12.7: A 'trapped dipole' uses RF traps to effectively disconnect parts of the antenna above the trap frequency

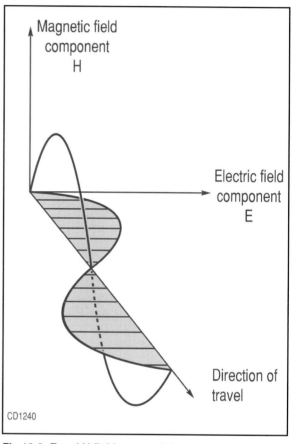

Fig 12.8: E and H fields are at right angles to each other and the direction of travel - The antenna element is parallel to the E field component axis

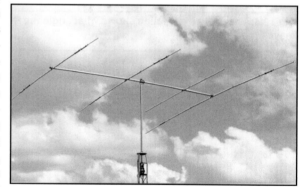

Picture 12.1: Cushcraft A-4S multiband HF beam

13: Propagation

Propagation is one of the most fascinating aspects of amateur radio. It is an aspect of the hobby that is still not fully understood, even by experts. However, there are some well-known general principles that you need to be aware of, and learning about propagation will give you an insight into what distances can be achieved and to help you get more out of the hobby.

Ground wave and sky wave

A HF radio signal can be received in two ways. It may travel to the receiver directly as a 'ground wave', or it can be refracted by the ionosphere as a 'sky wave' as shown in **Figure 13.1**. Amateurs will use the word **reflected** to freely describe this effect but you should remember the correct scientific term is **refracted**.

Ground wave

The ground wave signal travels directly to the receiver. The radio wave is gradually bent back down to the earth, and not bounced back like a ball from a wall. However, the ground wave signal is gradually absorbed by the ground it travels over and the distance it traverses is therefore limited. As frequency increases into the higher HF bands, the ground wave range is progressively reduced to just a few kilometres.

Sky wave

The sky wave signal is refracted by the ionosphere and returns to earth for reception. How the radio waves are propagated by the sky waves is quite complex and depends on the state of the ionosphere, the frequency in use and the angle at which the radio wave enters the ionosphere.

The Ionosphere

At this level we need to investigate the nature of the ionosphere in more detail. The ionosphere is composed of air, but it is much thinner than we are accustomed to at the earth's surface. Like other substances, air consists of molecules, which are the building blocks of matter. At the height of the ionosphere above the earth, the molecules are exposed to high levels of energetic radiation from the sun, causing the molecules to become 'ionised'.

The radiation that causes this ionisation is mostly of the type known as ultra-violet but it also includes solar particles. You don't need to know the scientific details of ionisation, just that it causes a reaction that splits up the molecules of air. In doing so it creates electrically charged particles called **ions** that refract radio signals.

You will recall that the ionosphere or layers of ionised gases, ranges from 70km to 400km in height. In reality, the ionosphere is not just one layer of ions in the atmosphere, but several, whose composition and height vary.

The layers are given letters, D, E and F, with F being the highest, as shown in **Figure 13.2**.

It is impossible to put exact figures on the heights of these layers as they are constantly changing. It is accepted that the D layer extends from about 60 to 90km and the E layer from about 100 to 125km. We need to think a little more carefully about the F layer. During the day when the earth faces the sun there are two F layers; F1 and F2 as shown in **Figure 13.3**.

We can also see that at night these two layers coalesce. During the day the height of the F1 layer is about 300km and the F2 layer is about 400km. At night the F layer is about 250 to 300km high.

Seasonal Variation in the Ionosphere

As a practising amateur you will have realised that contacts are more frequent in the summer than in the winter. Propagation is better in the summer because the ionosphere is much more active.

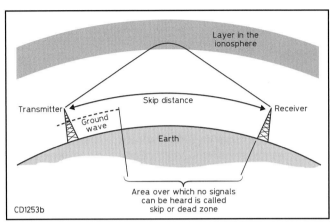

Fig 13.1: Illustrating skip distance – the total distance between transmitter and receiver

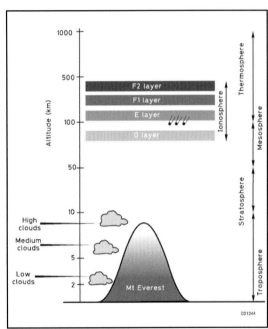

Fig 13.2: An overview of the ionosphere

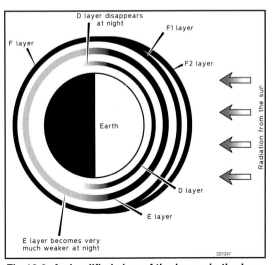

Fig 13.3: A simplified view of the layers in the Ionosphere over the period of the day

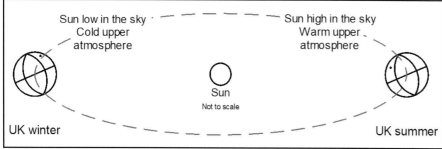

Fig 13.4 The relative positions of the Earth in its orbit around the sun

This *seasonal variation* can be explained by recalling that the axis of the Earth is not at a right angle to the plane of the orbit the Earth makes in its year long movement round the sun. It is tilted at an angle of 23.5° and we experience the different seasons of the year. See **Figure 13.4**.

If you were standing on the black dot, representing the UK, in the summer the sun would be high in the sky. The northern hemisphere will be receiving a lot of energy from the sun and the ionosphere will be very active. Six months later, during the winter, and standing on the same black dot, the sun will be low in the sky. The energy reaching the Earth is less and the ionosphere is less active.

There is, however, an additional fact we need to consider. During the winter months the ionosphere is more active during the mid-day hours. The radiation from the sun is continually producing ions but at the lower winter temperatures the rate at which they re-combine to give neutral atoms is slower. On balance, this means that during the middle of the day there is a higher concentration of ions and hence an increase in activity.

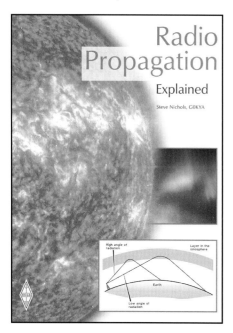

D layer

The D layer is the lowest in the atmosphere, where the molecules are closer together. When they are ionised they tend to absorb radio waves, rather than reflect them. The effect is quite dependent on frequency, and typically anything below 6MHz will not break through, limiting propagation to ground wave.

The lowest frequency which can pass through the D layer without significant absorption is known as the lowest usable frequency (LUF).

However, the D layer disappears quickly at night, when the sun's radiation is no longer present. This means that lower frequencies can then reach and be refracted from the F layer, and so longer distance contacts can be achieved. As a result, lower frequencies are suitable for long distance communications during evenings, with a preference towards the winter season.

E layer

The E layer appears to have little specific effect on HF propagation, but a form of propagation that takes place at VHF and the 24MHz and 28MHz upper HF bands is called 'sporadic-E'. As you might expect, it is associated with the E layer of the ionosphere. There are several theories about what causes Sporadic-E, but it is mostly a summertime phenomenon. It is very difficult to predict and can occur without warning, hence its name 'sporadic-E'.

Current theories point to sporadic-E being caused by wind shear and jet stream behaviour in lower weather systems affecting the E layer, although the exact mechanism is still not understood.

Contacts via sporadic-E often must be completed quickly, because the phenomena will not consistently reflect a given signal path. The lower VHF bands are most commonly refracted, usually up to around 150MHz. Because the E layer is lower than the F layer, signals will not be refracted quite so far, but multiple hops are possible. A single E layer hop is likely to yield a hop distance of up to 2000km.

F layer

The most important layers for HF communication are the F layers, F1, and F2, which are responsible for the refraction that makes most long-distance communication possible. As you might expect, because ionisation is caused by solar radiation, there is quite a big difference between day and night. The ions in the F layer recombine (to give neutral air molecules) more slowly with the onset of night. They are likely to remain present for several hours after sunset, continuing to provide refraction on both higher and lower frequencies. Solar radiation is at a maximum during the daytime, and so higher frequency HF bands are suitable for long distance communications during this time. There is also a seasonal variation, because of the lower solar radiation received during winter compared with summer.

Skip Distance and Skip Zone

The ability of the ionosphere to bend a radio wave back to earth depends on three factors, the degree of ionisation, the frequency of the wave and the angle at which the wave enters the ionosphere and can be summarised as:

- The higher the ionisation, the greater the bending ability.
- The higher the frequency the harder the wave is to bend.
- The shorter the distance between the transmitter and receiver the higher the angle the wave enters the ionosphere and the greater the bend required.

The angle at which the radio waves enter the ionosphere is also a major factor. If the waves strike at right angles, in other words travelling vertically from the antenna, it is likely that the waves will pass through the ionosphere, but as the angle reduces the waves start to bend as they enter the ionosphere. Eventually, the angle will be low enough that the wave will be bent back to return to earth some distance from the transmitting antenna. Once the refracted wave has returned to the surface of the earth it will be reflected back up to the ionosphere to start the second hop.

You should be clear in your mind about the meanings of refraction and reflection. Think of this as if you were dropping a stone vertically into a pond; you expect the stone to go straight to the bottom. This is the same as with a vertical wave hitting the ionosphere and going straight through. Now if you throw the same stone at a low angle it will skim across the surface with each bounce being part of a series of repeated refractions and reflections.

The frequency is important because as radio waves increase in frequency above about 15MHz they are less likely to be refracted. This maximum frequency that will be refracted over a particular path is known as the **maximum usable frequency (MUF)** and it changes as the ionosphere changes and explains why sometimes the higher HF bands are totally empty. There are tools available online which can report the MUF.

There is an area over which a radio signal will not be heard at all, because it is outside the range of the ground wave, and too close to receive the refracted sky wave. The area where no signals are heard is called the **skip zone** or, in some textbooks, the dead zone. Note that both these terms refer to the same zone. If you remember that, you are less likely to confuse the skip zone, where no signals can be heard, with the **skip distance**. The skip distance is the distance between the transmitting antenna and closest point the refracted signal returns to Earth. This is illustrated in Figure 13.1.

The HF bands are generally used for nationwide and international contacts, rather than local, cross-town QSOs – this is because cross-town contacts are most likely to be in the dead zone.

The furthest refractions are obtained from the F2 layer. The height of the layer and the curvature of the earth means that the longest hop is about 4000km. On a long path, perhaps involving more than one hop, it is possible that the MUF in the F-layer could be lower than the LUF set by absorption in the D-layer. Under those circumstances none of the HF bands can support communication. It is also worth remembering that the conditions in the ionosphere are set by the local time of day at the point the radio wave enters the ionospheric layer concerned. That need not be the same time as at the transmitter or receiver.

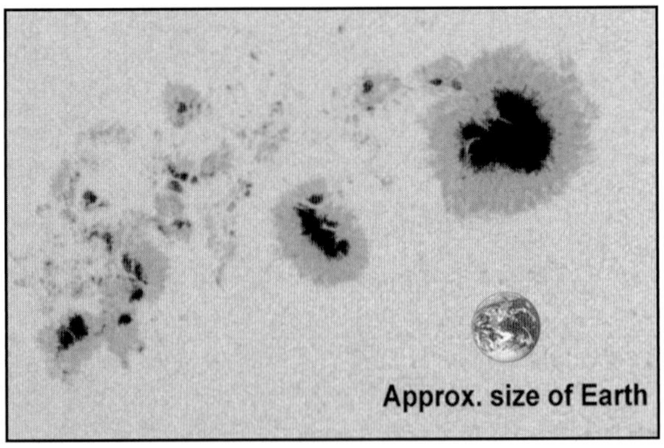

Fading

The path from the transmitter to the receiver via the ionosphere isn't a single line but several parallel paths each refracting of a slightly different bit off the ionosphere so that the lengths of each path differ slightly. It only requires a few metres, or tens of metres, difference for the various signals to arrive out of phase. These signals all add and subtract almost randomly so the level at the receiver input varies and can occasionally be so weak as to disappear. This effect is called fading, which can be fast, fluttering in a matter of seconds, or slow, coming and going over a few minutes. This phenomenon is also present on VHF bands and up.

Sunspots

For hundreds of years astronomers have recorded the number of sunspots on the sun's surface. The records show that there is a cycle of sunspot activity, with maximum numbers occurring every eleven years. Over the last one hundred years a clear link has been established between the number of sunspots and radio propagation. In a nutshell, an increase in the number of sunspots indicates an increase in the solar activity that causes ionisation in the ionosphere. Increased ionisation normally improves HF propagation.

Short path and long path

Worldwide propagation requires several hops. The shortest signal path (and therefore mostly likely the strongest signal path) follows a Great Circle path. Think of stretching a piece of elastic between two points on a globe of the earth. The shortest distance is part of a circle with its centre at the centre of the earth. However, there are still two routes to points almost opposite us on the globe, and the shortest path may not always result in the strongest signal.

New Zealand is the closest land mass opposite the UK on the earth. Consider the two paths between the two countries. The **short path** is the shorter distance round that circle (in this case, over the north pole), and the **long path** takes the other (longer) arc of the circle (over the south pole). At any time, half of the earth's surface is in sunlight and the other half in darkness, so it is quite possible that the long path may provide a stronger/clearer signal at certain times. Of course, this does not just apply to New Zealand, and these effects also depend on the local time of day at the various locations along the path.

VHF and UHF Propagation

As we go higher in frequency, signals are less likely to undergo ionospheric reflection. Several factors have an effect, including the angle at which the radio signals reach the ionosphere. As the frequency is increased, there exists a point where the radio waves are no longer reflected and pass directly through the ionosphere. This point is usually taken as 30MHz, the boundary between HF and VHF. However, in practice you will sometimes hear VHF signals being reflected from the ionosphere. Equally, there will be times when signals in the 28MHz band will not be reflected.

On VHF and UHF, day-to-day contacts can yield distances of around a few hundred kilometres for a modest setup on 144MHz. The ground terrain, temperature, air pressure and water vapor can cause some bending of radio waves, so ranges

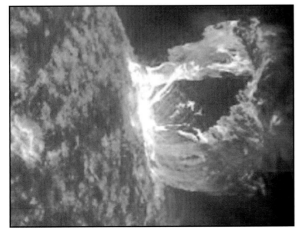

Picture 13.1: A closeup image of a Sunspot

Picture 13.2: Solar mass ejections interfere with radio signals on earth

may generally be a little further than you expect. VHF and UHF contacts tend to be more local than HF. On occasion, the troposphere can increase this range considerably. The troposphere is the layer of the atmosphere which is nearest the earth. Sometimes the troposphere can cause another type of bending of radio signals, under conditions known as a **temperature inversion**.

Under normal circumstances, the temperature of the air in the troposphere decreases with height. We all know that mountains become colder as we go higher. Sometimes an area of warmer air becomes trapped higher in the troposphere, along with some humidity. When this happens signals may become trapped in a layer (or duct) of the troposphere and return to earth at a distance. This phenomenon is known as **tropospheric ducting**. See **Figure 13.5**.

Tropospheric ducting can extend the range of VHF and UHF signals to many hundreds of kilometers. Because they are associated with patches of warmer air, temperature inversions occur more frequently in the summer.

You also need to be aware that the weather can reduce UHF propagation. Snow, hail and heavy rain contain lots of water and/or ice that absorb shorter wavelengths. Severe weather conditions can therefore prevent even local contacts on the UHF bands.

Typical Propagation

The following list gives a summary of the typical distances achievable per band:
- LF & MF: Ground wave during daytime can give national coverage (hundreds of km). It may be refracted at night to extend range and in the winter (thousands of km).
- HF: Ionospheric refraction gives typically reliable international communications (thousands of km), best during the day for frequencies above the LUF set by the D layer. The lower MUF means higher HF frequencies are poorer at night and in the winter.

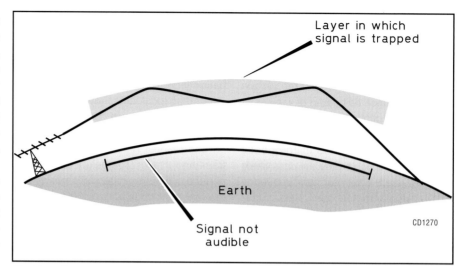

Fig 13.5: How a tropospheric duct can carry VHF signals across long distances

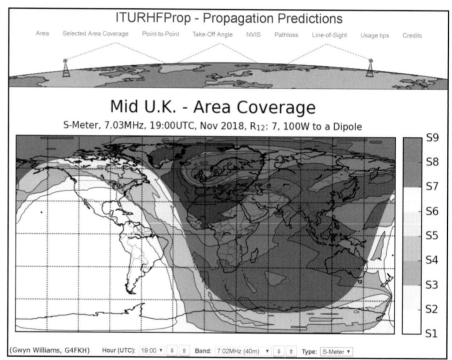

Fig 13.6: G4FKH's HF coverage modeling

There may be a winter peak of HF propagation around mid-day and early afternoon.
- VHF & UHF: Distances of around a few hundred kilometres, in normal conditions, can be enhanced by Sporadic-E (typically up to 2000 km) and tropospheric ducting (high-hundreds of km). UHF can be hindered by falling snow, hail and heavy rain.

At VHF and above, multipath propagation can occur where signals are reflected off objects, such as buildings and aircraft and the reflected signal is received in addition to the direct unreflected signal.

14: Measurements

Multi-meters
The multi-meter is an invaluable tool for the radio amateur, it allows for performing basic tests involving **voltage**, **current** and **resistance**, and with some effort can be used to test diodes and transistors.

Analogue & Digital Meters
You will probably encounter meters that have either an analogue – moving coil – display or a digital LCD or LED display. Both types are in common use and offer advantages and disadvantages.

When making measurements of a voltage or current that isn't stable, it's often easier to follow the movement with an analogue meter (**Picture 14.1**). These are also commonly used when adjusting circuits for a maximum or minimum reading. A disadvantage is that it can be difficult to read the meter scale with absolute precision.

A digital meter on the other hand is very easy to read and will quite often show several decimal places of precision, but the update rate may be slow, especially in cheaper meters, making it hard to follow a changing reading. Digital meters typically have a much higher input impedance, making the meter much less likely to affect the operation of the circuit under test.

Probes
Any multi-meter will have two coloured probes, black for negative or ground, and red for positive or live. The probes should be well insulated and have flexible leads. If you are using an old meter, or a very cheap one, make sure the probe insulation is good before you use them on any live circuitry.

The probes on a good meter are designed such that only a small tip of metal is exposed to make contact with the circuit. You must never touch the probe tips with your fingers when making a measurement; at best you may disturb the circuit under test with your body, at worst the circuit may be live.

You must also make sure you never allow the probes to touch each other when working on a circuit, as this will cause a short circuit, perhaps leading to circuit damage or worse if the circuit is mains powered and live.

Picture 14.1: An analogue multimeter

Making measurements
It is important that the meter is set to the correct measurement type and range before you connect it to a component or circuit, an incorrect setting can damage the meter and/or the item under test.

There are three measurements you need to know how to make for the Intermediate licence:
- Voltage or Potential Difference
- Current
- Resistance

Measuring Voltage
Voltage, potential difference, is always measured across, or in parallel, with a component in a circuit. Voltmeters have an extremely high resistance so as to have a negligible affect on the current in the test circuit. **Figure 14.1** shows where you would put the probes to measure the potential across resistors in a series circuit.

Measuring Current
Current is measured in Amps. Ammeters are always connected in series in a circuit. They can be inserted into the circuit at any point as the current is always constant. This is shown in **Figure 14.2**

A Consequence of using an Ammeter and Voltmeter
A voltmeter and ammeter will take electrical energy from the test circuit and affect the operation of the circuit. This is *not* considered at Intermediate Level and is deferred to the Full Licence.

Using a Voltmeter and Ammeter to measure Power.
There are occasions when we need to measure the power consumed by a load in a circuit.

For example, we might want to know the power used by a transceiver upon transmission.

It is common practice to represent the item drawing power as the **Load** as shown in **Figure 14.3** and you will recall from your

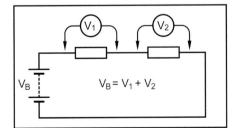

Fig 14.1: To measure voltage probes are placed across a component. The sum of the voltages across all the components in a series circuit is the same as the supply voltage

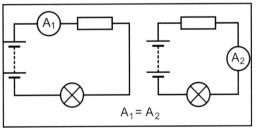

Fig 14.2: It does not matter where an Ammeter is connected in a series circuit. The current shown on meter A_1 is exactly the same as that shown on meter A_2

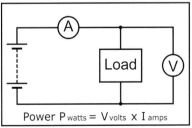

Fig 14.3: A circuit diagram for use of an Ammeter and Voltmeter to measure Power.

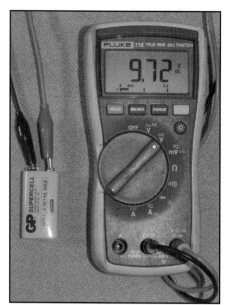

Picture 14.1: Checking the voltage of a 9V battery with a digital multimeter

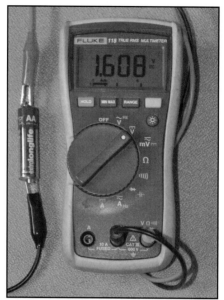

Picture 14.2: Checking the voltage of a 1.5 Volt AA cell

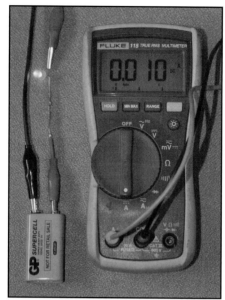

Picture 14.3: Measuring the current though an LED

Foundation Course that Power (Watts) = Amps x Volts

Always note steady readings on the ammeter and voltmeter to ensure accuracy. You should be aware that the power consumption will vary as to the mode you are using. There will be variations for AM and SSB. A constant power usage can be obtained by using a fixed audio tone or transmitting an FM signal. Be considerate to others and use a dummy load. More details and how to measure the RF output of a transmitter will be given in the Full Licence Course.

Using Multimeters

Voltage

A quick check of the condition of a battery is to check the voltage across its terminals and compare to its rated voltage. Take, for example, a standard 9 Volt battery. When fresh it should read somewhat over 9V, and when fully depleted will be around 8V.

To make this measurement we set the meter to the voltage setting and select a suitable range. Conversely, if the meter is auto-ranging, we can just allow the meter to do this for us. Connect the red and black probes to the positive and negative terminal of the battery. The meter should immediately show the voltage, and indeed we see a reading of 9.72 Volts in **Picture 14.2**.

If we now check a 1.5V AA cell with the meter fixed in this range, we will get a reading to two decimal places. However, if we adjust the range or allowed the meter to auto-range once again, we can see in Picture 14.2 that we get three decimal places for the voltage: 1.608 Volts.

With a manual ranged meter, we would set the range to the closest setting above the voltage to be measured. For example, with a 1.5 Volt AA cell, we may select 2V. If the voltage is unknown, you should start at the highest setting and work down to the lowest setting larger than the input voltage. This will give the more precise reading.

Starting with the voltage range higher than your expected voltage is generally best practice, as this will avoid damage to the meter, especially in the case of analogue meters where the needle may crash against the end-stop if the voltage is higher than the meter is set for.

Current

You know from your foundation studies that while potential difference (voltage) is measured **across** a component, current is measured **through** the component. To produce minimal disturbance to a circuit, the multi-meter, when set to measure current, will have a very low resistance between its probes. Effectively, the meter acts like a piece of wire. This is good for measuring current, but does mean you have to take care when positioning the probes. If you measure across a component with the meter in current setting, you effectively short-circuit that component, at best this will upset the operation of the circuit, at worst you may cause high current to flow though the meter and cause irreparable damage to both the meter and the circuit. Notice that the probe connections to the meter may be different to those when we measured voltage as in the pictures above.

In the circuit above, we expect around 10mA to flow though the LED. Since the meter shown is auto-ranging we simply set the meter to measure DC current, and the meter will automatically pick the most appropriate range. Had we been using a manual range meter, we would set the meter to a range larger than expected (always go a range higher, you can come down a range if needed once you confirm the current is low enough), and connect it in circuit. The meter shown displays a current of 0.010 Amps, which we know is the same as 10mA.

Resistance

Most multi-meters will offer both resistance and continuity measurement. The continuity measurement is useful for checking fuses, light bulbs and wires. The meter will produce a beep sound when the probes are connected to a very low resistance or short circuit, showing that there is a continuous electrical connection though the component under test.

To measure resistance the meter is first set to the resistance mode and the probes connected correctly to the meter (usually the same as for voltage measurements). The resistor to test is connected between the meter probes. If the meter is an auto-ranging type, the meter will display the resistance value to the best resolution it can achieve. If the meter requires manual ranging, it is common to start at the highest resistance range and work towards the lower resistance ranges. Analogue meters may require 'zeroing' first – this requires the probes be directly connected together and adjusting the meter to read zero.

15: The Examination

Is the online Intermediate exam like the Foundation?
Yes, the Intermediate exam is a multiple-choice assessment, with the question structures as in the Foundation exam. You will find a statement, question or diagram, with a choice of 4 answers. You should choose the correct or most appropriate answer from the 4.

You will be asked questions in line with the Intermediate syllabus and therefore you will require more knowledge than you did for the Foundation exam. The Intermediate exam has more questions: 46 to be exact, but you are allowed more time to complete it. You will find everything you need to know is covered within this book.

What will the questions be about?
The Intermediate exam is made up of 9 topics. Some of these topics will have more than one question (see Table 15.1). You will be tested on all topics, so make sure you have studied all the material within this book.

How can I prepare for the examination?
Everyone has their own method of revising for exams. Go back through all the chapters of this book, making notes and reminders in your book margin if that helps you. Alternatively, summarise key points on pieces of paper and collate them as your revision guide. Spend less time making notes on topics you understand and concentrate on the areas you are less comfortable with. The exercises suggested in Appendix 6 have been designed to help your understanding and learning. They are not part of the exam but you will find it very helpful if you can do them at home or as part of your club activities.

How should I tackle the examination?
It is very important to pace out your time and read each question very carefully. Do not jump ahead of yourself as this can result in you answering a question only to realise it was not what you expected and you have answered the wrong thing. So, read the questions and then read them again. It can be helpful to work out the approximate time per question and use this to set your pace – use the time you gain on easier questions to spend longer answering questions you find more difficult.

What happens if I don't score enough?
If you do not pass the examination, do not become disheartened, and whatever you do, don't give up! You will need to arrange another test. A good approach would be to use this time to revise, paying attention to the topics that caused you difficulties before.

Licence Conditions	6
Technical Aspects	14
Transmitters and Receivers	7
Feeders and Antennas	4
Propagation	3
EMC	4
Operating Practices and Procedures	2
Safety	3
Measurements and Construction	3

Table 15.1: Approximate breakdown of Intermediate examination questions

Appendix 1
The relationship between power and voltage

Why do we say that at the half power point the Voltage is 0.707 x its Initial Value? There are two methods of explaining this.

Method 1
P = I V

And, as we know that

$$I = \frac{V}{R}$$

it follows that

$$P = \frac{V^2}{R}$$

Rearranging we get,

$$R = \frac{V^2}{P}$$

The resistance of the circuit is constant (it does not change when we reduce power) and we can see that,

$$\frac{V^2_{Half\ Power}}{P_{Half\ Power}} = \frac{V^2_{Full\ Power}}{P_{Full\ Power}}$$

Rearranging we get,

$$\frac{V^2_{Half\ Power}}{V^2_{Full\ Power}} = \frac{P_{Half\ Power}}{P_{Full\ Power}}$$

And as

$$\frac{V^2_{Half\ Power}}{V^2_{Full\ Power}} = \frac{1}{2} = 0.5$$

$$\frac{V_{Half\ Power}}{V_{Full\ Power}} = \sqrt{0.5}$$

$$\frac{V_{Half\ Power}}{V_{Full\ Power}} = 0.707$$

This tells us that at half power the voltage is 0.707 times its original value. Hence, if the Initial Voltage was 1 Volt, the half power voltage is 0.707 Volts

Method 2
You may prefer an alternative explanation, Starting from

$$P = \frac{V^2}{R}$$

R is a constant so in determining the relationship between P and V we can ignore it.

If we wish to double the power then V^2 must double.

It follows that V must increase by √2 because (√2)² is, obviously, 2

Similarly to halve the power, V^2 must halve meaning that V must reduce by 1/√2 which is 1/1.414 or 0.707.

The above is not required learning for the Intermediate Exam. It is included for those who like to see rigorous proof.

Appendix 2
Protective Multiple Earthing

With certain types of mains electricity supply, particularly a type known as TNC-s, a type of Protective Multiple Earthing practice (PME), there are fault conditions, albeit unlikely to occur, where the mains house earth can rise considerably above true earth potential. Ordinarily this is not a serious problem because the entire house is at that potential and it requires a potential difference to cause an electric shock.

Introducing true earth by having an RF earth in the shack will provide that potential difference. The normal solution is to cross-bond the two earths with a substantial conductor. Substantial because it may become the neutral side return path for all the connected appliances in the house. It must be stressed that such a procedure is not always appropriate and it is a notifiable addition to the house electrical system in the same way as a new wiring circuit is notifiable.

Since services such as water and gas may be delivered by plastic pipes it may be that pipes in the house, including central heating radiators are no longer effectively earthed by default. Consequently there are circumstances where it may not be appropriate for them to be bonded to the house safety earth at the electricity supply Main Earth Terminal.

These are circumstances where it is essential to get proper professional advice from a qualified tradesperson. Any changes to the earthing arrangements, including the provision of an RF earth is notifiable under part-P of the building regulations. The qualified tradesperson will be familiar with this requirement and can advise accordingly. Hearsay, however well meaning, from fellow amateurs will not do. Their exact circumstances may well be different.

You can download from the EMC pages of the RSGB website a booklet that details common methods of earthing domestic supplies:

EMC07-Earthing and the Radio Amateur

Appendix 3
Scientific Notation

Scientific notation is a way of writing very large and very small numbers in a much simpler way. Sometimes it is referred to as using "powers of ten" and can be explained with some simple examples.

Some examples of expressing numbers as powers of ten.

300   is   3 x 100   or 3×10^2
3000   is   3 x 1000   or 3×10^3
3,000,000 is   3 x 1,000,000   or 3×10^6

0.3   is $3 \times \frac{1}{10}$ or 3×10^{-1}

0.03   is $3 \times \frac{1}{100}$ or 3×10^{-2}

0.003   is $3 \times \frac{1}{1000}$ or 3×10^{-3}

0.0003   is $3 \times \frac{1}{10,000}$ or 3×10^{-4}

0.00003 is $3 \times \frac{1}{100,000}$ or 3×10^{-5}

Multiplication Example
What is the value of 200 x 6000?

This is the same as
$(2 \times 10^2) \times (6 \times 10^3) = 2 \times 6 \times 10^2 \times 10^3$

$2 \times 6 = 12$ and $10^2 \times 10^3 = (10 \times 10) \times (10 \times 10 \times 10) = 10^5$

Answer is 12×10^5

We can take this further, 12 is 1.2 x 10

So 12×10^5 can be rewritten as

$1.2 \times 10 \times 10^5$

which is the same as

1.2×10^6

The superscripted numbers are called *indices*. You will have noticed that in multiplication *the indices are added together,*

for $1.2 \times 10^1 \times 10^2 \times 10^3$

Number	Arithmetic presentation	Can be written as,	Can be said as,
10	10	10^1	times ten
100	10 x 10	10^2	times ten squared
1,000	10 x 10 x10	10^3	times ten cubed
10,000	10 x 10 x10 x 10	10^4	times ten to the power of four
100,000	10 x 10 x10 x 10 x10	10^5	times ten to the power of five
1,000,000	10 x 10 x10 x 10 x10 x 10	10^6	times ten to the power of six

Table A3.1: Numbers bigger than 1

Number	Arithmetic presentation	Can be written as,	Can be said as,
0.1	$\frac{1}{10}$	10^{-1}	times ten to the power of -1
0.01	$\frac{1}{100}$	10^{-2}	times ten to the power of -2
0.001	$\frac{1}{1000}$	10^{-3}	times ten to the power of -3
0.0001	$\frac{1}{10,000}$	10^{-4}	times ten to the power of -4
0.00001	$\frac{1}{100,000}$	10^{-5}	times ten to the power of -5
0.000001	$\frac{1}{1,000,000}$	10^{-6}	times ten to the power of -6

Table A3.2: Numbers smaller than 1

we have
$1 + 2 + 3 = 6$,
hence 1.2×10^6

Division Example
What is the value of 12000 divided by 20?

or $\frac{12,000}{20}$

This is the same as

$\frac{1.2 \times 10^4}{2 \times 10^1} = 0.6 \times \frac{10^4}{10^1}$

Because this is a *division we subtract* the bottom index from the upper index, $4 - 1 = 3$

$0.6 \times \frac{10^4}{10^1}$

becomes $0.6 \times 10^3 = 6 \times 10^2 = 600$

Four Amateur Radio Examples
1. What is the current in a circuit where the potential difference is 1 millivolt and the resistance is 2 kΩ? Give your answer in µAmps

   1mV is 0.001V which is 1×10^{-3} Volts and 1kΩ is 1000Ω, hence 2kΩ is 2×10^3 Ω

   Current = $\frac{Volts}{Ohms} = \frac{1 \times 10^{-3}}{2 \times 10^3}$

   = $0.5 \times 10^{-6} = 5 \times 10^{-7}$ Amps
   (We subtract the indices; -3 -3 = -6)
   1Amp is 10^6 µAmps
   Current = $5 \times 10^{-7} \times 10^6 = 0.5$ µAmps

2. What is the frequency of an electromagnetic wave with a wavelength of 1.5cm?

Speed of electromagnetic radiation is
3 X 10^8 m/s and 1.5cm is 1.5 x 10^{-2}m
Speed of light = frequency x wavelength
Re-arranging the formula we have
Frequency

$$= \frac{speed\ of\ light}{wavelength} = \frac{3 \times 10^8}{1.5 \times 10^{-2}}$$

= 2 x 10^{10} = 20 x 10^9 or 20GHz
(We subtract the indices; 8-(-2) =10)

3. The antenna voltage at the input of a transceiver of 50Ω impedance is measured and found to be 100μVolts. What power does this represent?

$$Power = \frac{V^2}{R}$$

100 μVolts is 100 x 10^{-6}V which is 10^{-4}V

Hence,
(100 x 10^{-6})^2 becomes (10^{-4})^2
which is the same as 10^{-8}

$$Power = \frac{10^{-8}}{50} = 0.02 \times 10^{-8}$$

which is the same as 2 x 10^{-10} Watt
Power = 2 x 10^{-10} Watt
A Watt is 10^9 nanoWatts
Hence,
Power = 2 x 10^{-10} x 10^9 = 0.2nW

4. What is the half power bandwidth of a 14.15 Mz signal in a tuned circuit with a Quality Factor, Q, of 100? Give your answer in both MHz and kHz.
Quality Factor

$$= \frac{frequency\ of\ signal}{Bandwidth}$$

$$100 = \frac{14.15 \times 10^6}{Bandwidth}$$

Rearranging we have,
Bandwidth

$$= \frac{14.15 \times 10^6}{100}$$

= 0.1415 x 10^{6}Hz or 0.1415MHz
1MHz is 1000kHz or 10^{3}kHz
Bandwidth = 0.1415 x 10^3 or 141.5kHz

Note: Modern calculators will allow you to enter numbers as powers of ten but you should remember that calculators vary in the way they operate and in the exam you should use the calculator you are most familiar with. It certainly helps to practice a few examples.

Appendix 4
Formulae from RSGB EX308

Ohm's Law V= IR	Power P=V×I
Series R$_T$= R$_1$+R$_2$+R$_3$	Parallel $\frac{1}{R_T} = \frac{1}{R_1} + \frac{1}{R_2} + \frac{1}{R_3}$
Potential divider $V_{out} = V_{in} \frac{R_2}{R_1+R_2}$	
Series $\frac{1}{C_T} = \frac{1}{C_1} + \frac{1}{C_2} + \frac{1}{C_3}$	Parallel C$_T$= C$_1$+C$_2$+C$_3$
Series L$_T$= L$_1$+L$_2$+L$_3$	Parallel $\frac{1}{L_T} = \frac{1}{L_1} + \frac{1}{L_2}$
AC $V_{rms} = \frac{V_{peak}}{\sqrt{2}}$	AC $t = \frac{1}{f}$ $f = \frac{1}{t}$
Inductor X$_L$= 2πfL	Capacitor $X_C = \frac{1}{2\pi fC}$
Tuned circuit $Q = \frac{f_C}{f_U-f_L} = \frac{centre\ frequency}{bandwidth}$	
Transformer $V_s = V_p \frac{N_s}{N_p}$	Transformer $I_p = I_s \frac{N_s}{N_p}$
Transistor I$_C$= βI$_B$	
Velocity of radio waves in free space v= 3×10^8 m/s = 300,000,000 m/s	Frequency & wavelength v= fλ
antenna erp= power × gain (linear)	

You are provided with a copy of EX308 when you take your examination which includes the formulae shown. These formulae are for you to use when answering some questions, so, do not need to be memorised.

Appendix 5: Experimentation for Learning

There is much that can be gained from experimentation and clubs and individual students are strongly advised to work through the following examples at their own pace. Hopefully, this will happen within your club's normal training activities and you will have the benefit of discussion with experienced practitioners to advance your skills and understanding.

Project kits

There are many kits available targeted specifically at radio amateurs trying soldering for the first time. Most good radio kits include many hints and tips on good soldering practice.

Building a DC circuit and making some measurements

This section will take you through building a simple DC circuit. You will explore the effect of resistors on the current though the circuit and the brightness of the bulb.

You may use any construction techniques you like, however the one shown here is simple, very easy to assemble and works well.

Materials and components

- A small test board - approximately 100mm square or larger, e.g. pin board, soft wood, balsa wood, non-silver side of a cake base, cork board
- 2 x small bulb holders
- 2 x Bulbs – 2.5 Volt, 0.1 Amp
- Resistor (R1) – 10Ω
- Resistor (R2) – 47Ω
- Metal pins or small drawing pins – 10 required (must be brass or copper)
- Battery holder with 'snap' connector for two AA cells in series
- Battery connecting clip (preferably with red and black wires fitted as standard)
- AA cells – 2 required
- 2 x switches or paperclips to make switches from.
- Connecting wire – 20cm of single stranded insulated wire
- Connecting wire – 20cm of stranded wire coiled up for fly leads.
- Soldering equipment – soldering iron, stand, 20cm of multicore solder
- Tools – long-nosed pliers, wire strippers, wire cutters

The Circuit

The schematic diagram of the circuit is shown here in **Figure A6.1**; you might notice that there is a switch across R2 – this is not a mistake.

Construction

Using **Picture A6.1** as a guide:

1. Place pins on the test board in the approximate pattern shown, but not too far apart. Do not push them fully home. If you're using paperclips as switches trap one loop of the paperclip under the pins as shown. The clips can then be slid to make or break the contact.
2. Fix the bulb holder to the board with double sided adhesive tape or suitable screws. If the bulbs you are using have wires attached, solder the wires to the pins as shown

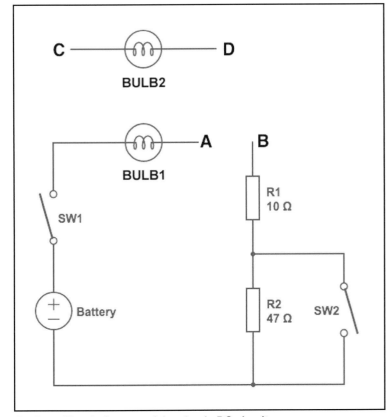

Fig A6.1: Circuit diagram of the simple DC circuit

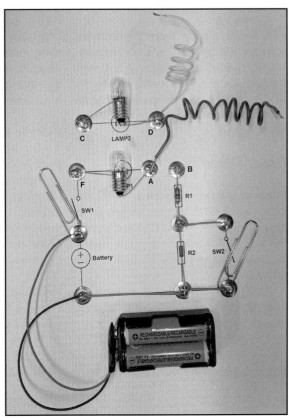

Picture A6.1: layout of the simple DC circuit

3. Strip insulation from both ends of three short lengths of wire (each about 3-4cm in length) and from the ends of the battery clip wires.
4. Solder the two resistors onto the pins in the positions shown. Ensure the heads of the pins are clean before attempting to solder to them (e.g. by rubbing them with wire wool or abrasive paper). You may need to trap some of the leads under the pins to prevent the first resistor falling off as the second is soldered on.
5. WARNING! The pins get very hot and the solder can take a while to cool. Wait for it to do so, or you may burn yourself. Do not blow on the solder to cool it.
6. Solder a connecting wire between the negative battery pin and the bottom of R2. Make sure the joints are nice and shiny.
7. Solder another wire from the bottom of R2 to the switch SW2.
8. Solder a wire from the top of R2, where it joins R2 to the other pin for switch SW2.
9. Solder the red (positive) wire of the battery to the pin for switch SW1.
10. Solder the black (negative) wire of the battery to the negative battery pin.
11. Solder the one fly-lead to point A.
12. Solder the second fly-lead to point D.

Check your connections, make sure switch SW2 is open, insert the batteries correctly and then turn bulb 1 on, by closing switch SW1 and shorting A to B with the fly-lead; do not solder the wire to pin B, just touch it with the bare wire. If all is fine, bulb-1 will glow rather dimly and bulb-2 will be off.

Experiments

Effect of resistors in series

With switch SW1 closed, switch SW2 open and points A and B joined by the fly-lead, bulb-1 will glow rather dimly. You may have to shade the bulb to see any glow.

The circuit has the two resistors R1 and R2 in series, limiting the current though the bulb from the batteries.

Using Picture A6.1 as an example, if you close SW2, bulb-1 should become noticeably brighter. You have shorted out R2, effectively removed it from the circuit, allowing a larger current to flow, so the bulb glows more brightly.

Using Ohm's law

Keeping the circuit as above with bulb-1 glowing more brightly, use a volt meter to measure the voltage across R1. Connect the positive lead of your volt meter to point B and the negative side of your volt meter to the other side of R1. This will measure the potential difference across R1.

Now with SW1 and SW2 still both closed, disconnect the fly-lead between points A and B. Instead, connect an ammeter between points A and B, so you can measure the current through the circuit when just bulb-1 and R1 are connected in series. Connect the positive lead of your current meter to point A and the negative lead to point B. The bulb should illuminate at the same brightness as before.

You can then use Ohm's law to calculate the resistance of R1. Remembering Ohm's Law:

$$V = I \times R$$

We can rearrange this using the magic triangle we saw earlier to calculate R:

$$R = \frac{V}{I}$$

You should expect to find a value of resistance close to 10Ω, assuming the measurements and circuit construction have been undertaken correctly. Many modern meters have the ability to measure resistance directly, so you can also check your work using a resistance measurement.

Bulbs in series

If you now move the fly-lead to connect point A to point C and use the second fly-lead to connect point D to point B, you will see both bulbs glow, but dimmer than just bulb-1 on its own. You have put both bulbs in series, the combined resistance of the bulbs limits the current, so the bulbs do not glow as brightly.

Further experiments

Can you figure out how to connect the bulbs in parallel? (Hint – you may need to add some extra wires) What do you see?

What effect does opening switch SW-2 have when the bulbs are in series and parallel?

You may wish to measure the current in the circuit, you can do this with a multi-meter set to current range. If you open switch SW-1 you can connect the meter with the positive lead to the battery positive pin, and the negative meter lead to point F. The meter will now read the current in the circuit as you experiment.

A transistor as a switch

This section will guide you through building a simple transistor switch circuit. You will make measurements of the components used as you build it and use those to calculate the gain of the transistor in the circuit.

The transistor switch relies on the gain of the transistor to allow a small base current to control a larger collector current.

Materials and components
- A small test board - approximately 100mm square or larger, e.g. pin board, soft wood, balsa wood, non-silver side of a cake base, cork board
- 1 x small bulb holder
- 1 x Bulb – 2.5 Volt, 0.1 Amp
- Resistor (R1) – 50Ω
- Resistor (R2) –3.3kΩ
- 1 x NPN Transistor. BFY50 or BFY51 works well.
- Metal pins or small drawing pins – 7 required (must be brass or copper)
- Battery holder with 'snap' connector for 3 AA cells in series
- Battery connecting clip (preferably with red and black wires fitted as standard)
- AA cells – 3 required
- Connecting wire – 20cm of single stranded insulated wire
- Soldering equipment – soldering iron, stand, 20cm of multicore solder
- Tools – long-nosed pliers, wire strippers, wire cutters

Construction

Before soldering down the components you need to measure the value of the resistors and note them down, these values will be used in the transistor gain calculation later.

Using **Picture A6.2** as a guide build the circuit in the same way as you did the previous simple DC circuit, taking care with soldering. When you come to fitting the transistor, it must be placed correctly. If you are using the BFY50 / BFY51 device suggested, the emitter is the lead nearest the tag on the case, the base is the middle lead and the collector is the other lead.

When you connect the battery, the lamp should glow. If it does not, disconnect the battery and check your connections, as the most likely cause is the transistor connections. Double check these and try again.

Experiments

With the battery connected and the lamp glowing, use the multi-meter on the voltage range to measure the voltage across points **A** and **B**, this is the battery voltage under load. Note this down.

Using the meter again, measure the voltage across points **A** and **C** and points **A** and **D**. Note down these readings.

The gain of a transistor is defined to be the ratio of the collector current to the base current, the formula is:

$$\beta = \frac{I_c}{I_b}$$

Where I_c is the collector current and I_b is the base current.

You could break the circuit and insert an ammeter to measure the base and collector currents, but we can also calculate I_c and I_b from the measurements you've already made.

Remembering Ohm's Law

$$V = I \times R$$

You have measured R when you built the circuit, and you have measured V when the circuit works, so we can calculate the current though each of the resistors and into the transistor.

The collector current, I_c is equal to the current though R_1, which is equal to the voltage across points **A** and **D** divided by the measured value of R_1.

$$I_c = \frac{V}{R_1}$$

The base current, I_b is equal to the current though R_2, which is equal to the voltage across points **A** and **C** divided by the measured value of R_2.

$$I_b = \frac{V}{R_2}$$

The transistor gain is then

$$\beta = \frac{I_c}{I_b}$$

The gain of a BFY50 / BFY51 transistor should be around 40. If your answer is wildly different from this, check you have measured the resistor values correctly. You must measure them before you connect them to the circuit. You could also check the transistor you are using as transistor types other than BFY50 / BFY51 will have different gains.

Experiments with oscillators

This section will lead you through some of the experiments and measurements around oscillators.

You will need a crystal oscillator and a LC oscillator, your local radio club may be able to help you with these. You will also need a general coverage receiver or a spectrum analyser and a low pass filter.

Using a general coverage receiver for oscillator testing

The most readily accessible calibration tool is a receiver which can cover the range of the VFO under test. As well as being a useful means of listening to the amateur bands, a general coverage receiver is an excellent item of test equipment.

Even the best screened VFO will radiate a low-level signal and as receivers are designed to pick up very weak signals it should be possible to listen to the VFO and measure its frequency on the receiver's display or dial.

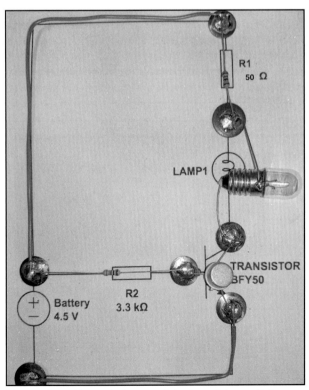

Picture A6.2: Example layout of the transistor switch

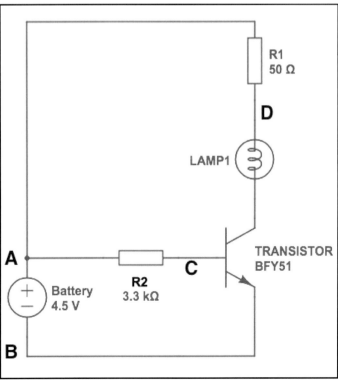

Fig A6.2: Circuit diagram of a simple transistor switch

A5: Experimentation for Learning

Firstly, select the approximate frequency on the receiver and set the mode to CW or SSB. You need to tune to the VFO signal until the pitch reduces and stop when you cannot hear the VFO anymore. This is known as 'zero beat', the point where there is no difference between the VFO and the receiver's frequency. But how do we know if the receiver is showing the right frequency?

The receiver itself can be calibrated against a signal of known accuracy, such as the standard frequency services (e.g. WWV from the USA on 2500, 5000, 10000, 15000 and 20000kHz), or by a simple item of test equipment known as a 'crystal calibrator'. Some receivers have a crystal calibrator built into them. If you use one of these methods and find that your receiver is constantly 1kHz out, you will know to make suitable corrections when calibrating your VFO.

Stability

To demonstrate that a crystal oscillator is stable when subjected to reasonable temperature changes and mechanical shock, you should first determine the frequency the crystal oscillator is oscillating on and tune the receiver or spectrum analyser to the same frequency. You shouldn't hear significant drift (change in tone in the radio) or see any movement of the trace on a spectrum analyser. You can tap the oscillator case or warm it gently, there should be little to no frequency change.

To demonstrate that a variable frequency (LC) oscillator is not very stable when subjected to reasonable temperature changes and mechanical shock, the above test can be repeated as with the crystal oscillator on the LC oscillator. You'll find the frequency isn't nearly as stable, as the tone in the radio will drift or the trace on a spectrum analyser will move.

Harmonics

To find (at least) the 2nd and 3rd harmonics from an RF oscillator, you can use either a receiver or spectrum analyser. Using the crystal oscillator, tune the radio to twice the oscillator frequency, you should be able to hear a tone in the radio. The same should be true at three times the oscillator frequency. These are the second and third harmonics of the fundamental oscillator frequency. If you're using a spectrum analyser set to a wide range, you may be able to see these harmonics as peaks in the trace at 2 times and 3 times the fundamental frequency. You may even see peaks at other harmonic frequencies depending on the oscillator type.

Once you have found the harmonics using either the radio or the spectrum analyser, you should try connecting a low pass filter to the output of the oscillator. You should notice that the higher harmonics are now reduced in strength or removed altogether

Calibrating a VFO

Some interesting work can be done with Variable Frequency Oscillators (VFOs). Your club or fellow amateurs may have one you can borrow or alternatively you may care to make one. An example VFO you can build yourself can be found on the Books Extra page of the RSGB website www.rsgb.org/booksextra

To calibrate a variable RF frequency oscillator you will need a receiver, frequency counter or spectrum analyser of known accuracy. Band edges and two intermediate points need to be marked at zero beat. This section will take you through calibrating a VFO.

VFO Adjustment

The precise calibration adjustments will depend on the VFO you are using, but there are normally two types of adjustments that can be made.

Firstly, the inductor in the VFO's tuned circuit may have a core on a screw thread that can be moved in and out of the coil. This is normally used to set the VFO's mid-range. By setting the variable capacitor to its mid position, the inductor's core can be adjusted whilst listening to the middle of the intended frequency range. Special (non-metallic) adjustment tools are made for use with inductor cores. Using an ordinary screwdriver will not only affect the oscillator's frequency but it *could* shatter the core and render it useless. If you do not have a suitable tool you would be well advised to buy one from your inductor supplier or an online store or auction site.

The second adjustment is normally a pre-set variable capacitor or trimmer that limits the range of the main variable capacitor. Its adjustment may require some trial and error, swinging the VFO through its range, checking the frequencies at each end then adjusting the trimmer to increase or decrease the swing.

Ideally the VFO should only operate within the amateur band edges but it is quite acceptable for it to cover a wider frequency range, providing the dial is clearly calibrated to show the limits of the amateur frequency allocation.

Fitting RF connectors

You will need to be able to fit an RF connector of a suitable type to a piece of coaxial cable. The connector may be solder, crimp or compression type, or a combination of these. It is a good opportunity to make a coaxial cable patch lead, for connecting from your radio to SWR meter or AMU.

During your Foundation training you will have met a couple of RF connectors; the PL259 and the BNC. You may already have used them as part of your Foundation station, but you might not have fitted

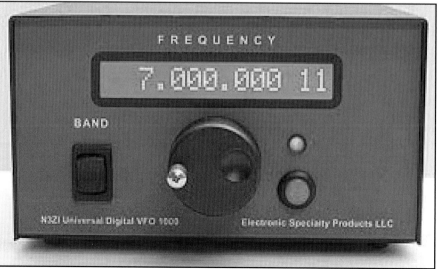

Picture A6.3: A digital VFO

them yourself.

If your equipment uses 'N' type connectors it would be sensible for you to use one of those.

Details of fitting various RF plugs can be found in many of the books available on antennas, but details of the compression-type PL259 and BNC are set out here. Compression N connectors follow identical fitting instructions as the BNC. You may also find it helpful to observe others fitting a plug, and another amateur or YouTube video may help.

The PL259

There are several types of PL259. The instructions below are for a plug that uses a compression washer to ensure good electrical contact with the braid and to hold the coaxial cable firmly in the plug. These are widely recognised as being good quality, reliable connectors and are available for most popular sizes of amateur radio coax. The exact fitting process may be slightly different for your plugs but these instructions should give you the basic idea of the procedure. Often connectors come with a small slip of paper detailing specific instructions or measurements for the connector in question.

1. Check that you have all the components of a complete plug: locking bush, compression washer, flange, and main body (see **Picture A6.4 (a)**).
2. If you are using a piece of coax that already has a plug on one end, or a long length of coax, thread the locking bush and the compression washer on and push them down the cable for about 15-20cm. If you are using a short piece of coax with nothing on the other end, you can do this later.
3. Prepare the end of the coax by removing about 50mm of the outer covering. The best way to do this is with a cable stripper. If you use a sharp knife instead, run the blade around the cable, but don't cut too deep (into the fine strands of the braid) as they can break off and cause short circuits inside the plug.
4. Using a small screwdriver, tease out the strands of the braid right back to the outer covering.
5. Slide the flange over the coax inner until the narrow sleeve is inserted between the braid and the coax inner.
6. Using a pair of sharp scissors, trim the braid back so that it is level with the flange (see Picture A6.4 (b)).
7. Cut the inner insulation about 1-2mm away from the flange and pull it off. Be very careful not to cut into the inner conductor. If the inner is damaged you will need to chop the coax back and start all over again.
8. Some coax cables have a single heavier gauge wire for the inner conductor and if yours is one of them, move on to the next step. If your inner conductor has multiple strands of thinner wire, twist them together and lightly tin the end to prevent any of the strands bending back when you get to the next step.
9. Guide the inner conductor through the body of the plug until it appears out of the end of the pin, checking that all the strands have come through. Keep pushing it until the flange is right up against the main body of the plug.
10. Push the compression washer up to the flange and push the locking bush up to the plug body.
11. Screw the body of the plug onto the locking bush and tighten it with two pairs of pliers or suitable spanners – you should find the main body of the plug has two flats to provide a good grip.
12. Solder the inner conductor to the pin. Make sure that you do not get too much solder on the inside, as it can flow up the pin and short the pin to the braid. You may find it helpful to hold the plug in a vice, a wooden clothes peg or a pair of pliers with a rubber band around the handles, with the pin facing slightly downwards to prevent solder flowing into the main body of the plug. You are looking for a small pool of solder on the tip of the pin with the inner conductor in the centre well soldered.
13. Once the solder has cooled, trim off any excess inner conductor and, if required, use a small file to remove

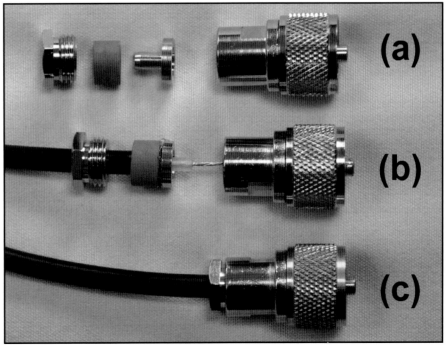

Picture A6.4: A compression-type PL259
(a) shows the parts of the plug
(b) shows the part assembled plug. Note how the braid has been trimmed around the flange
(c) shows the completed plug ready for testing

A5: Experimentation for Learning

any excess solder that has made the pin too big to fit an SO239 socket (see Picture A6.4 (c)). Be careful not to file any contact plating off.

14. Using a multi-meter set to a high resistance range, check that there is no short circuit between the body of the plug and its pin. Also check for low resistance readings between the plug pin and the inner conductor at the other end of the coax, and between the plug body and the braid at the other end of the coax. If all is well, you can use the coax and plug. If not, chop it off and start again!

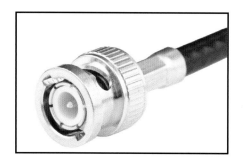

The BNC

As with the PL259, there are several variations on the basic design of the compression BNC. These instructions cover one type. You may have to make some changes depending on your actual plug. Notice that there are different types of plug made by different manufacturers. You can see two common types in **Figure A6.3**.

1. Check that you have all the components of the plug (see Figure A6.3).
2. Thread the nut, the metal washer and the rubber washer over the coax.
3. Prepare the end of the coax by removing about 30mm of the outer covering, using the same technique as for the PL259.
4. Thread the cone over the braid up to the outer covering.
5. Using a small screwdriver, tease out the strands of the braid right back to the cone.
6. Trim the braid back to about 10mm from the cone using a pair of sharp scissors or good quality wire cutters and then fold them back over the cone. Try to spread them evenly around the cone (see Figure A6.3).
7. Trim the braid so it just covers the cone.
8. Cut the inner insulation exactly 5mm away from the braid/cone and pull it off. Be very careful not to cut into the inner conductor. If the inner is damaged, you will need to chop the coax back and start all over again.
9. If the inner conductor is stranded (some have a single, heavier gauge wire), twist them together, but *do not* tin the wire.
10. Cut the inner conductor exactly 4mm from the inner insulation.
11. Holding the coax in a vice, a wooden clothes peg or a pair of pliers with a rubber band around the handles, fit the pin over the inner conductor and solder it in place. This is best done by heating the pin and feeding the solder through the hole in the side of the pin. Allow the pin to cool for a minute or so. Once the pin has cooled, pull on the pin to check the solder joint is good.
12. Place the body of the plug over the pin, being careful not to disturb the braid on the cone. Satisfy yourself that the pin has seated correctly in the insulation and is protruding from the front of the plug.
13. Gently push the rubber washer up against the cone and then the metal washer against the rubber and screw the nut into the body of the plug.
14. Tighten the nut with two spanners or pairs of pliers as with the PL259.
15. Using a multi-meter set to a high resistance range, check that there is no short circuit between the body of the plug and its pin, and that there is good continuity between the plug pin and the inner conductor at the other end of the coax, and between the plug body and the braid at the other end of the coax. If all is well, you can use the coax and plug. If not, chop it off and start again!

A correctly connected and terminated coaxial cable ensures that the RF field only exists within the cable and is not affected by objects outside the cable.

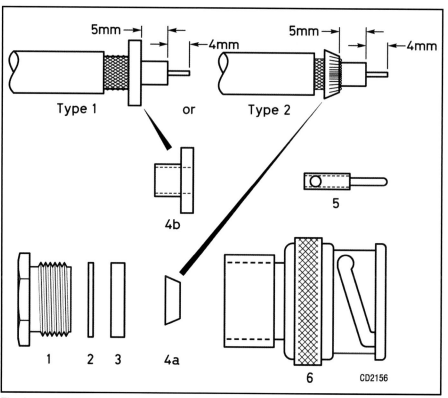

Fig A6.3: Exploded view of a BNC plug. Note that differences exist between plug types from various manufacturers

LINDARS RADIOS

"A Modern Company With Old Fashioned Values"

WE ALWAYS HAVE A LARGE SELECTION OF NEW & USED AMATEUR RADIO EQUIPMENT

STARTER PACKAGES OFFERED FOR THE NEWLY LICENCED* (Subject to Stock Availability)

HF TRANSCEIVERS
VHF / UHF TRANSCEIVERS
SWR / POWER METERS, POWER SUPPLYS, ATU, AERIALS, COAX
PATCH LEADS, * BOOKS

USED AMATEUR RADIO EQUIPMENT PURCHASED AND SOLD

www.AmateurRadioSales.co.uk

Call Justin Lindars 01935 474265 with your requirements or for just good plane advice

Email: lindarsradio@gmail.com
2 Buckland Rd, Pen Mill Trading Est,
Yeovil BA21 5EA

PayPal VISA MasterCard SWITCH

Shop Opening Times: Tuesday - Friday 9.30am - 2.30pm, Saturday 9.30am - 2.30pm

Would you like a noise free radio?
Get a bhi DSP noise cancelling product!

bhi

NES10-2 MK3
Amplified DSP speaker:
- Rotary filter select switch
- 8 DSP filter levels 9 to 35dB
- 5W input & 2.7W audio out
- 3.5 mm mono headphone socket
- On/off audio bypass switch
- 12 to 24VDC (500mA)

Simply plug in the audio and connect the power!

ParaPro EQ20 DSP with parametric equaliser
- 20W audio power amplifier with parametric equalisation
- Fine tune your audio
- Optional bhi DSP Noise Cancelling and Bluetooth connectivity
- 4 versions available: EQ20, EQ20-DSP, EQ20-B, EQ20B-DSP
- Can be used with your bhi Dual In-Line, NEIM1031 & Compact In-line units

Shape your receive audio to suit your ears!

Dual In-Line

Mono/stereo DSP noise eliminating module
Latest bhi DSP noise cancelling
- 8 Filter levels 8 to 40dB - 3.5mm Mono or stereo inputs - Line level in/out - 7 watts mono speaker output - Headphone socket - Easy to adjust and setup - Ideal for DXing, club stations, special event stations and field day events - Supplied boxed with user manual and audio/power leads - Suitable for use with many radios and receivers including Elecraft K3, KX3 & FlexRadio products

Testimonial
'the on air performance in improving readability of weak SSB signals or those in noisy conditions were excellent'
RadCom

Compact In-Line
Compact handheld mono/stereo in-line DSP noise cancelling unit
- Easy to use rotary controls for all functions - New improved DSP noise cancelling - Use with mono or stereo inputs - 8 filter levels 9 to 35dB - Ideal for portable use & DXing - Use with headphones or a small speaker
- 12V DC power or 2 x AA batteries
- Over 40 hours battery life
- Size: 121mm x 70mm x 33mm
- Suitable for use with Elecraft K3 & KX3

DSPKR
10W amplified DSP noise cancelling speaker
- Easy control of DSP filter
- 7 filter levels
- Sleep mode
- Filter select & store function
- Separate volume control
- Input overload LED
- Headphone socket
- Supplied with user manual and fused DC power lead

DESKTOP
- 10W amplified DSP noise cancelling base station speaker
- Rotary volume and filter level controls
- 8 filter levels 9 to 35dB
- Speaker level and line level audio inputs
- 3.5mm Headphone socket
- Size 200(H)x150(D) x160(W)mm, Weight 1.9 Kg
- For use with most radios, receivers & SDR including Elecraft & FlexRadio

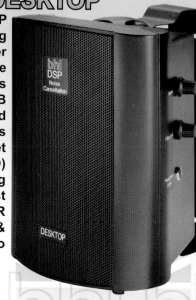

bhi Ltd, 22 Woolven Close
Burgess Hill, RH15 9RR, UK Tel: 01444 870333 www.bhi-ltd.com